UNIVERSITY OF

LR/LEND/001

KU-365-502

Rapid Prototyping

Pri

App

EC 2014

WITHDRAWN

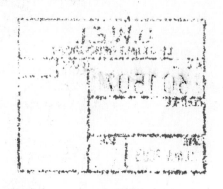

WITHDRAWN

Rapid | Second Edition |
Prototyping

Principles and
Applications

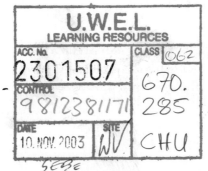

U.W.E.L.
LEARNING RESOURCES

ACC. No.
2301507

CLASS 062

CONTROL
9812381171

670.
285

DATE
10. NOV. 2003

SITE
WV

CHU

CHUA C. K., LEONG K. F. and LIM C. S.

Nanyang Technological University, Singapore

World Scientific
New Jersey • London • Singapore • Hong Kong

Published by

World Scientific Publishing Co. Pte. Ltd.

5 Toh Tuck Link, Singapore 596224

USA office: Suite 202, 1060 Main Street, River Edge, NJ 07661

UK office: 57 Shelton Street, Covent Garden, London WC2H 9HE

British Library Cataloguing-in-Publication Data

A catalogue record for this book is available from the British Library.

RAPID PROTOTYPING: PRINCIPLES AND APPLICATIONS, 2nd Edition

Copyright © 2003 by World Scientific Publishing Co. Pte. Ltd.

All rights reserved. This book, or parts thereof, may not be reproduced in any form or by any means, electronic or mechanical, including photocopying, recording or any information storage and retrieval system now known or to be invented, without written permission from the Publisher.

For photocopying of material in this volume, please pay a copying fee through the Copyright Clearance Center, Inc., 222 Rosewood Drive, Danvers, MA 01923, USA. In this case permission to photocopy is not required from the publisher.

ISBN 981-238-117-1

ISBN 981-238-120-1 (pbk)

Printed by FuIsland Offset Printing (S) Pte Ltd, Singapore

My wife, Wendy, and children, Cherie, Clement and Cavell, whose forbearance, support and motivation have made it possible for us to finish this book.

Chee Kai

Soi Lin, for her patience and support, and Qian who brings us cheer and joy.

Kah Fai

My wife, Eugenia, and family for their support and motivation.

Chu Sing

CONTENTS

FOREWORD

When I heard Chua Chee Kai, Leong Kah Fai and Lim Chu Sing were planning a new edition of this book, it put a smile on my face. The reason is that the rapid prototyping (RP) industry is in desperate need of up-to-date and comprehensive educational products. A growing number of colleges and universities are adding course work and projects on RP, but making the experience a good one for students can be a challenge without quality textbooks and coursework. *Rapid Prototyping: Principles and Applications* not only aims to satisfy this need, it delivers.

The book is filled with practical information that even experienced users will find it helpful. The chapters include detailed descriptions of the available RP processes, giving readers an excellent overview of what's available commercially. The photographs, illustrations, and tables make these chapters visually appealing and straightforward. References and problems at the end of each chapter help readers expand their understanding of the topics presented.

The chapter on data formats provides detail that new and advanced users alike will appreciate. It dives into related formats, such as IGES and SLC, but its in-depth coverage of the STL file format, including its limitations, problems, and solutions, makes this chapter shine.

The book's sections on RP applications explain why and how organizations are putting RP to work. Examples range from building flight-ready metal castings at Bell Helicopter to producing a human skull to aid in brain surgery at Keio University Hospital in Japan. The new information on tooling expands the reader's understanding of how organizations of all types are applying this powerful technology.

The depth and breadth presented in this book make it crystal clear that Chua, Leong and Lim have a strong understanding of RP technologies and applications. The information is written and presented in an easy-to-follow format, and the book's accompanying CD is icing on the cake.

This book sits in a special place on my bookshelf. If you don't have it, get it. Thank you, Professors Chua, Leong and Lim, for producing such an outstanding book and for allowing me to share my thoughts on it.

Terry Wohlers
President
Wohlers Associates, Inc.

PREFACE

The focus on productivity has been one of the main concerns of industries worldwide since the early 1990s. To increase productivity, industry has attempted to apply more computerized automation in manufacturing. Amongst the latest technologies to take the industry by storm is *Rapid Prototyping Technologies*, otherwise also known as *Solid Freeform Fabrication, Desktop Manufacturing* or *Layer Manufacturing Technologies*.

The revolutionary change in factory production techniques and management requires a direct involvement of computer-controlled systems in the entire production process. Every operation in this factory, from product design, to manufacturing, to assembly and product inspection, is monitored and controlled by computers. CAD/CAM or Computer-Aided Design and Manufacturing has emerged since the 1960s to support product design. Until the mid-1980s, it has never been easy to derive a physical prototype model, despite the existence of CNC or Computer Numerical Controlled Machine Tools. Rapid Prototyping Technologies provide that bridge from product conceptualization to product realization in a reasonably fast manner, without the fuss of NC programming, jigs and fixtures.

With this exciting promise, the industry and academia have internationally established research centers for Rapid Prototyping (RP), with the objectives of working in this leading edge technology, as well as of educating and training more engineers in the field of RP. An appropriate textbook is therefore required as the basis for the development of a curriculum in RP. The purpose of this book is to provide an introduction to the fundamental principles and application areas in RP. The book traces the development of RP in the arena of Advanced Manufacturing Technologies and explains the principles underlying each of the RP techniques. It also covered the RP processes, their specifications and their 2002 estimated prices. In this second edition, twelve more new RP techniques are introduced, bringing the total number to 30 RP techniques. The book would not be complete without emphasizing the importance of RP applications in manufacturing and other industries. In addition to industrial examples provided for the vendor, an entire chapter is devoted to application areas.

One key inclusion in this book is the use of multimedia to enhance understanding of the technique. In the accompanying compact disc (CD), animation is used to demonstrate the working principles of major RP techniques such as Stereolithography, Solid Ground Curing, Laminated Object Manufacturing, Fused Deposition Modeling, Selective Laser Sintering and Three-Dimensional Printing.

In addition, the book focuses on some of the very important issues facing Rapid Prototyping today, and these include, but are not limited to:

(1) The problems with the *de facto* STL format.
(2) The range of applications for tooling and manufacturing, including biomedical engineering.
(3) The benchmarking methodology in selecting an appropriate RP technique.

The material in this book has been used and revised several times for professional courses conducted for both academia and industry audiences since 1991. Certain materials were borne out of research conducted in the School of Mechanical and Production Engineering at the Nanyang Technological University, Singapore. To be used more effectively for graduate or final year (senior year) undergraduate students in Mechanical and Production Engineering or Manufacturing Engineering, problems have been included in this textbook version. For university professors and other tertiary-level lecturers, the subject Rapid

Prototyping can be combined easily with other topics such as: CAD, CAM, Machine Tool Technologies and Industrial Design.

Chua C. K.
Associate Professor

Leong K. F.
Associate Professor

Lim C. S.
Associate Professor

School of Mechanical and Production Engineering
Nanyang Technological University
50 Nanyang Avenue
Singapore 639798

ACKNOWLEDGMENTS

Firstly, we would like to thank God for granting us His strength throughout the writing of this book. Secondly, we are grateful to our spouses, Wendy, Soi Lin and Eugenia, and our children, Cherie, Clement, Cavell Chua and Leong QianYu for their patience, support and encouragement throughout the year it took to complete this book.

We wish to thank Professor Yue Chee Yoon, Dean, School of Mechanical and Production Engineering (MPE), Nanyang Technological University (NTU), Singapore for his unwavering support of this project. We would also like to acknowledge the valuable support from the administration of NTU. In addition, we would like to thank our current and former students Chan Lick Khor, Chong Fook Tien, Chow Lai May, Foo Hui Ping, Kwok Yew Heng, Angeline Lau Mei Ling, Phung Wen Jia, Tang Ho Hwa, Toh Choon Han and Wee Kuei Koon, our colleague Mr Lee Kiam Gam for their valuable contributions to the multimedia CD which demonstrates some major RP techniques. We would also like to express sincere appreciation to special assistants Deborah Cheah Meng Lin and Alex Tan Kok Wai for their help and effort in the coordinating and getting it published on time.

Much of the research work which have been published in various journals and now incorporated into some chapters in the book can be attributed to our former students, Tong Mei, Micheal Ko, Simon Cheong, William Ng, Chong Lee Lee, Lim Bee Hwa, Chua Ghim Siong, Ko Kian Seng, Chow Kin Yean, Wong Yeow Kong, Althea Chua, Terry Chong, Frank Lin Sin Ching, Tan Yew Kwang, Chiang Wei Meng, Ang Ker Ser, Ng Chee Chin and Melvin Ng Boon Keong, and colleagues Professor Robert Gay, Mr Lee Han Boon, Dr Du Zhaohui and Dr Cheah Chi Mun.

We would also like to extend our special appreciation to Mr Terry Wohlers for his foreword and granting the permission to use his executive summary of the Wohler's report, Dr Ming Leu of the University of Missouri-Rolla, USA, Dr Amba D Bhatt, National Institute of Standards and Technology, Gaithersburg, MD, USA, Dr Philip Coane and Dr Goettert, Institute for Micromanufacturing, Ruston, LA, USA, and Prof Fritz Prinz of Stanford University, USA for their assistance.

The acknowledgment would not be complete without mentioning the contribution of the following companies, in the order in which they appear in the book, for supplying information about their products:

(1) 3D Systems Inc., USA
(2) Cubital Ltd., Israel
(3) D-MEC Ltd., Japan
(4) CMET Inc., Japan
(5) Autostrade Co. Ltd., Japan
(6) Teijin Seiki Co. Ltd., Japan
(7) Meiko Co. Ltd., Japan
(8) Cubic Technologies Inc., USA
(9) Stratasys Inc., USA
(10) Kira Corp. Ltd., Japan
(11) Solidscape Inc., USA
(12) Beijing Yinhua Rapid Prototypes Making and Technology Co. Ltd., China
(13) CAM-LEM Inc., USA
(14) Ennex Corp., USA
(15) EOS GmbH, Germany
(16) Z Corp., USA
(17) Optomec Inc., USA
(18) Soligen Inc., USA
(19) Fraunhofer-Gessellschaft, Germany
(20) Acram AB, Sweden
(21) Aeromet Corp., USA
(22) Generis GmbH, Germany

(23)   Therics Inc., USA

(24)   Extrude Hone Corp., USA

Last but not least, we also wish to express our thanks and apologies to the many others not mentioned above for their contributions to the success of the book. We would appreciate your comments and suggestions on this second edition book.

Chua C. K.
Associate Professor

Leong K. F.
Associate Professor

Lim C. S.
Associate Professor

ABOUT THE AUTHORS

CHUA Chee Kai is an Associate Professor and the Head of the Systems and Engineering Management Division at the School of Mechanical and Production Engineering, Nanyang Technological University, Singapore. Dr Chua has extensive teaching and consulting experience in Rapid Prototyping (RP) including being the advisor to RP bureau start-ups. His research interests include the development of new techniques for RP as well as their applications in industry. He was the keynote speaker for conferences on "Trends of RP" and has sat on several programme committees for international RP conferences. Dr Chua sits on the Editorial Advisory Board of Rapid Prototyping Journal, MCB Press, UK. He is the author of over 140 publications, including books, book chapters, international journals and conference proceedings and has one patent in his name.

Dr Chua can be contacted by email at mckchua@ntu.edu.sg or visit his webpage at http://www.ntu.edu.sg/MPE/Divisions/systems/Faculty/ChuaCheeKai.htm.

LEONG Kah Fai is an Associate Professor at the School of Mechanical and Production Engineering, Nanyang Technological University, Singapore. He is the Programme Director for the Master of Science Course in Smart Product Design and was the Founding Director of the Design Research Centre at the University. He graduated from the National University of Singapore and Stanford University for his undergraduate and postgraduate degrees, respectively. He is actively involved in local design, manufacturing and standards work. He has chaired several committees in the Singapore Institute of Standards and

Industrial Research and Productivity Standards Board, receiving Merit and Distinguished Awards for his services in 1994 and 1997, respectively. He has authored over 60 publications, including books, book chapters, international journals, and conferences.

Dr Leong can be contacted by email at mkfleong@ntu.edu.sg or visit his webpage at http://www.ntu.edu.sg/MPE/Divisions/mechatronics &design/Faculty/LeongKahFai.htm.

LIM Chu Sing is an Associate Professor at the School of Mechanical and Production Engineering, Nanyang Technological University, Singapore. He is concurrently the Deputy Director of the university's Biomedical Engineering Research Centre. Dr Lim received his BEng (Hons) and PhD from Loughborough University in 1992 and 1995, respectively. He obtained the National Science & Technology Board (NSTB) Postdoctoral Fellowship and the Tan Chin Tuan Fellowship to serve at the NTU and Duke University, respectively. He received several awards for his work including the Japanese Chamber of Commerce & Industry (Singapore Foundation) Education Award. Dr Lim serves in several professional and advisory committees. He has served as a consultant to several international and local companies. Dr Lim is the author of over 40 journal and conference publications, one book chapter and has two patents in his name.

Dr Lim can be contacted by email at mchslim@ntu.edu.sg or visit his webpage at http://www.ntu.edu.sg/MPE/Divisions/systems/Faculty/ LimChuSing.htm.

LIST OF ABBREVIATIONS

3D	=	Three-Dimensional
3DP	=	Three-Dimensional Printing
ABS	=	Acrylonitrile Butadiene Styrene
AIM	=	ACES Injection Molding
BPM	=	Ballistic Particle Manufacturing
CAD	=	Computer-Aided Design
CAE	=	Computer-Aided Engineering
CAM	=	Computer-Aided Manufacturing
CBC	=	Chemically Bonded Ceramics
CD	=	Compact Disc
CIM	=	Computer-Integrated Manufacturing
CLI	=	Common Layer Interface
CMM	=	Coordinate Measuring Machine
CNC	=	Computer Numerical Control
CSG	=	Constructive Solid Geometry
CT	=	Computerized Tomography
DMD	=	Direct Metal Deposition
DMLS	=	Direct Metal Laser Sintering
DSP	=	Digital Signal Processor
DSPC	=	Direct Shell Production Casting
EBM	=	Electron Beam Melting
EDM	=	Electric Discharge Machining
FDM	=	Fused Deposition Modeling
FEA	=	Finite Element Analysis
FEM	=	Finite Element Method
GPS	=	Global Positioning System
HPGL	=	Hewlett-Packard Graphics Language

IGES	=	Initial Graphics Exchange Specification
LAN	=	Local Area Network
LCD	=	Liquid Crystal Display
LEAF	=	Layer Exchange ASCII Format
LED	=	Light Emitting Diode
LENS	=	Laser Engineered Net Shaping
LMT	=	Layer Manufacturing Technologies
LOM	=	Laminated Object Manufacturing
M-RPM	=	Multi-Functional RPM
MEM	=	Melted Extrusion Modeling
MJM	=	Multi-Jet Modeling System
MJS	=	Multiphase Jet Solidification
MRI	=	Magnetic Resonance Imaging
NASA	=	National Aeronautical and Space Administration
NC	=	Numerical Control
PC	=	Personal Computer
PCB	=	Printed Circuit Board
PDA	=	Personal Digital Assistant
PLT	=	Paper Lamination Technology
RFP	=	Rapid Freeze Prototyping
RP	=	Rapid Prototyping
RPI	=	Rapid Prototyping Interface
RPM	=	Rapid Prototyping and Manufacturing
RPS	=	Rapid Prototyping Systems
RPT	=	Rapid Prototyping Technologies
RSP	=	Rapid Solidification Process
SAHP	=	Selective Adhesive and Hot Press
SCS	=	Solid Creation System
SFF	=	Solid Freeform Fabrication
SFM	=	Solid Freeform Manufacturing
SGC	=	Solid Ground Curing
SLA	=	StereoLithography Apparatus
SLC	=	StereoLithography Contour
SLS	=	Selective Laser Sintering

SSM	=	Slicing Solid Manufacturing
SOUP	=	Solid Object Ultraviolet-Laser Printing
STL	=	StereoLithography File
UV	=	Ultraviolet

Chapter 1
INTRODUCTION

The competition in the world market for manufactured products has intensified tremendously in recent years. It has become important, if not vital, for new products to reach the market as early as possible, before the competitors [1]. To bring products to the market swiftly, many of the processes involved in the design, test, manufacture and market of the products have been squeezed, both in terms of time and material resources. The efficient use of such valuable resources calls for new tools and approaches in dealing with them, and many of these tools and approaches have evolved. They are mainly technology-driven, usually involving the computer. This is mainly a result of the rapid development and advancement in such technologies over the last few decades.

In product development [2], time pressure has been a major factor in determining the direction of the development and success of new methodologies and technologies for enhancing its performance. These also have a direct impact on the age-old practice of prototyping in the product development process. This book will introduce and examine, in a clear and detailed way, one such development, namely, that of Rapid Prototyping (RP).

1.1 PROTOTYPE FUNDAMENTALS

1.1.1 Definition of a Prototype

A prototype is an important and vital part of the product development process. In any design practice, the word "prototype" is often not far from the things that the designers will be involved in. In most dictionaries, it is defined as a noun, e.g. the Oxford Advanced Learner's Dictionary of Current English [3] defines it as (see Figure 1.1):

1

> *A prototype is the first or original example of something that has been or will be copied or developed; it is a model or preliminary version; e.g.: A prototype supersonic aircraft.*

Figure 1.1: A general definition of a prototype

However, in design, it often means more than just an artefact. It has often been used as a verb, e.g. prototype an engine design for engineering evaluation, or as an adjective, e.g. build a prototype printed circuit board (PCB). To be general enough to be able to cover all aspects of the meaning of the word prototype for use in design, it is very loosely defined here as:

> *An approximation of a product (or system) or its components in some form for a definite purpose in its implementation.*

This very general definition departs from the usual accepted concept of the prototype being physical. It covers all kinds of prototypes used in the product development process, including objects like mathematical models, pencil sketches, foam models, and of course the functional physical approximation of the product. Prototyping is the process of realizing these prototypes. Here, the process can range from just an execution of a computer program to the actual building of a functional prototype.

1.1.2 Types of Prototypes

The general definition of the prototype contains three aspects of interests:

(1) the implementation of the prototype; from the entire product (or system) itself to its sub-assemblies and components,

(2) the form of the prototype; from a virtual prototype to a physical prototype, and

(3) the degree of the approximation of the prototype; from a very rough representation to an exact replication of the product.

The implementation aspect of the prototype covers the range of prototyping the complete product (or system) to prototyping part of, or a sub-assembly or a component of the product. The complete prototype, as its name suggests, models most, if not all, the characteristics of the product. It is usually implemented full-scale as well as being fully functional. One example of such prototype is one that is given to a group of carefully selected people with special interest, often called a focus group, to examine and identify outstanding problems before the product is committed to its final design. On the other hand, there are prototypes that are needed to study or investigate special problems associated with one component, sub-assemblies or simply a particular concept of the product that requires close attention. An example of such a prototype is a test platform that is used to find the comfortable rest angles of an office chair that will reduce the risk of spinal injuries after prolonged sitting on such a chair. Most of the time, sub-assemblies and components are tested in conjunction with some kind of test rigs or experimental platform.

The second aspect of the form of the prototype takes into account how the prototype is being implemented. On one end, virtual prototypes that refers to prototypes that are nontangible, usually represented in some form other than physical, e.g. mathematical model of a control system [4]. Such prototypes are usually studied and analyzed. The conclusions drawn are purely based upon the assumed principles or science that has been understood up to that point in time. An example is the visualization of airflow over an aircraft wing to ascertain lift and drag on the wing during supersonic flight. Such prototype is often used when either the physical prototype is too large and therefore takes too long to build, or the building of such a prototype is exorbitantly expensive. The main drawback of these kinds of prototypes is that they are based on current understanding and thus they will not be able to predict any unexpected phenomenon. It is very poor or totally unsuitable for solving unanticipated problems. The physical model, on the other hand, is the tangible manifestation of the product, usually built for testing and experimentation. Examples of such prototypes include a mock-up of a cellular telephone that looks and feels very much like the real product but without its intended functions. Such a prototype may be used purely for aesthetic and human factors evaluation.

The third aspect covers the degree of approximation or representativeness of the prototype. On one hand, the model can be a very rough representation of the intended product, like a foam model, used primarily to study the general form and enveloping dimensions of the product in its initial stage of development. Some rough prototypes may not even look like the final product, but are used to test and study certain problems of the product development. An example of this is the building of catches with different material to find the right "clicking" sound for a cassette player door. On the other hand, the prototype can be an exact full scale exact replication of the product that models every aspects of the product, e.g. the pre-production prototype that is used not only to satisfy customer needs evaluation but also addressing manufacturing issues and concerns. Such "exact" prototypes are especially important towards the end-stage of the product development process.

Figure 1.2 shows the various kinds of prototypes placed over the three aspects of describing the prototype. Each of the three axes represents one aspect of the description of the prototype. Note that this illustration is not meant to provide an exact scale to describe a prototype, but serves to demonstrate that prototypes can be described along these three aspects.

Rapid prototyping typically falls in the range of a physical prototype, usually are fairly accurate and can be implemented on a component level or at a system level. This is shown as the shaded volume shown in Figure 1.2. The versatility and range of different prototypes, from complete systems to individual components, that can be produced by RP at varying degrees of approximation makes it an important tool for prototyping in the product development process. Adding the major advantage of speed in delivery, it has become an important component in the prototyping arsenal not to be ignored.

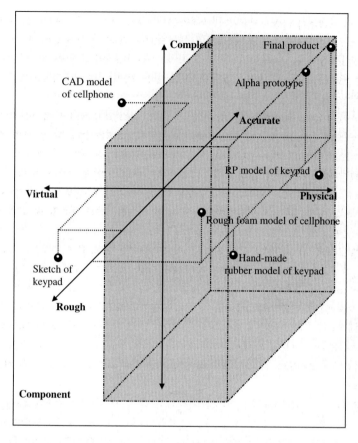

Figure 1.2: Types of prototypes described along the three aspects
of implementation, form and approximation

1.1.3 Roles of the Prototypes

The roles that prototypes play in the product development process are
several. They include the following:

(1) Experimentation and learning
(2) Testing and proofing
(3) Communication and interaction
(4) Synthesis and integration
(5) Scheduling and markers

To the product development team, prototypes can be used to help the thinking, planning, experimenting and learning processes whilst designing the product. Questions and doubts regarding certain issues of the design can be addressed by building and studying the prototype. For example, in designing the appropriate elbow-support of an office chair, several physical prototypes of such elbow supports can be built to learn about the "feel" of the elbow support when performing typical tasks on the office chair.

Prototypes can also be used for testing and proofing of ideas and concepts relating to the development of the product. For example, in the early design of folding reading glasses for the elderly, concepts and ideas of folding mechanism can be tested by building rough physical prototypes to test and prove these ideas to see if they work as intended.

The prototype also serves the purpose of communicating information and demonstrating ideas, not just within the product development team, but also to management and client (whether in-house or external). Nothing is clearer for explanation or communication of an idea than a physical prototype where the intended audience can have the full experience of the visual and tactile feel of the product. A three-dimensional representation is often more superior than that of a two-dimensional sketch of the product. For example, a physical prototype of a cellular phone can be presented to carefully selected customers. Customers can handle and experiment with the phone and give feedback to the development team on the features of and interactions with the phone, thus providing valuable information for the team to improve its design.

A prototype can also be used to synthesize the entire product concept by bringing the various components and sub-assemblies together to ensure that they will work together. This will greatly help in the integration of the product and surface any problems that are related to putting the product together. An example is a complete or comprehensive functional prototype of personal digital assistant (PDA). When putting the prototype together, all aspects of the design, including manufacturing and assembly issues will have to be addressed, thus enabling the different functional members of the product development team to understand the various problems associated with putting the product together.

Prototyping also serves to help in the scheduling of the product development process and is usually used as markers for the end or start of the various phases of the development effort. Each prototype usually marks a completion of a particular development phase, and with proper planning, the development schedule can be enforced. Typically in many companies, the continuation of a development project often hinges on the success of the prototypes to provide impetus to management to forge ahead with it.

It should be noted that in many companies, prototypes do not necessary serve all these roles concurrently, but they are certainly a necessity in any product development project.

The prototypes created with Rapid Prototyping technologies will serve most if not all of these roles. Being accurate physical prototypes that can be built with speed, many of these roles can be accomplished quickly and effectively, and together with other productivity tools, e.g. CAD, repeatedly with precision.

1.2 HISTORICAL DEVELOPMENT

The development of Rapid Prototyping is closely tied in with the development of applications of computers in the industry. The declining cost of computers, especially of personal and mini computers, has changed the way a factory works. The increase in the use of computers has spurred the advancement in many computer-related areas including Computer-Aided Design (CAD), Computer-Aided Manufacturing (CAM) and Computer Numerical Control (CNC) machine tools. In particular, the emergence of RP systems could not have been possible without the existence of CAD. However, from careful examinations of the numerous RP systems in existence today, it can be easily deduced that other than CAD, many other technologies and advancements in other fields such as manufacturing systems and materials have also been crucial in the development of RP systems. Table 1.1 traces the historical development of relevant technologies related to RP from the estimated date of inception.

Table 1.1: Historical development of Rapid Prototyping and related technologies

Year of Inception	Technology
1770	Mechanization [4]
1946	First Computer
1952	First Numerical Control (NC) Machine Tool
1960	First commercial Laser [5]
1961	First commercial Robot
1963	First Interactive Graphics System (early version of Computer-Aided Design) [6]
1988	First commercial Rapid Prototyping System

1.2.1 Three Phases of Development Leading to Rapid Prototyping

Prototyping or model making in the traditional sense is an age-old practice. The intention of having a *physical* prototype is to realize the conceptualization of a design. Thus, a prototype is usually required before the start of the full production of the product. The fabrication of prototypes is experimented in many forms — material removal, castings, moulds, joining with adhesives etc. and with many material types — aluminium, zinc, urethanes, wood, etc.

Prototyping processes have gone through three phases of development, the last two of which have emerged only in the last 20 years [7]. Like the modeling process in computer graphics [8], the prototyping of physical models is growing through its third phase. Parallels between the computer modeling process and prototyping process can be drawn as seen in Table 1.2. The three phases are described as follows.

1.2.2 First Phase: Manual Prototyping

Prototyping had began as early as humans began to develop tools to help them live. However, prototyping as applied to products in what is

Table 1.2: Parallels between geometric modeling and prototyping

Geometric Modeling	Prototyping
❶ **First Phase: 2D Wireframe**	❶ **First Phase: Manual Prototyping**

• Started in mid-1960s
• Few straight lines on display may be:
  • circuit path on a PCB
  • plan view of a mechanical component
• "Natural" drafting technique
	• Traditional practice for many centuries
	• Prototyping as a skilled crafts is:
	  • traditional and manual
	  • based on material of prototype
	• "Natural" prototyping technique

❷ **Second Phase: 3D Curve and Surface Modeling**	❷ **Second Phase: Soft or Virtual Prototyping**
• Mid-1970s • Increasing complexity • Representing more information about precise shape, size and surface contour of parts	• Mid-1970s • Increasing complexity • Virtual prototype can be stressed, simulated and tested, with exact mechanical and other properties
❸ **Third Phase: Solid Modeling**	❸ **Third Phase: Rapid Prototyping**
• Early 1980s • Edges, surfaces and holes are knitted together to form a cohesive whole • Computer can determine the inside of an object from the outside. Perhaps, more importantly, it can trace across the object and readily find all intersecting surfaces and edges • No longer ambiguous but exact	• Mid-1980s • Benefit of a hard prototype made in a very short turnaround time is its main strong point (relies on CAD modeling) • Hard prototype can also be used for limited testing • Prototype can also assist in the manufacturing of the products

considered to be the first phase of prototype development began several centuries ago. In this early phase, prototypes typically are not very sophisticated and fabrication of prototypes takes on average about four weeks, depending on the level of complexity and representativeness [9]. The techniques used in making these prototypes tend to be craft-based and are usually extremely labor intensive.

1.2.3   Second Phase: Soft or Virtual Prototyping

As application of CAD/CAE/CAM become more widespread, the early 1980s saw the evolution of the second phase of prototyping — *Soft or Virtual Prototyping*. Virtual prototyping takes on a new meaning as more computer tools become available — computer models can now be stressed, tested, analyzed and modified as if they were physical prototypes. For example, analysis of stress and strain can be accurately predicted on the product because of the ability to specify exact material attributes and properties. With such tools on the computer, several iterations of designs can be easily carried out by changing the parameters of the computer models.

Also, products and as such prototypes tend to become relatively more complex — about twice the complexity as before [9]. Correspondingly, the time required to make the physical model tends to increase tremendously to about that of 16 weeks as building of physical prototypes is still dependent on craft-based methods though introduction of better precision machines like CNC machines helps.

Even with the advent of Rapid Prototyping in the third phase, there is still strong support for virtual prototyping. Lee [10] argues that there are still unavoidable limitations with rapid prototyping. These include material limitations (either because of expense or through the use of materials dissimilar to that of the intended part), the inability to perform endless what-if scenarios and the likelihood that little or no reliable data can be gathered from the rapid prototype to perform finite element analysis (FEA). Specifically in the application of kinematic/dynamic analysis, he described a program which can assign physical properties of many different materials, such as steel, ice, plastic, clay or any custom material imaginable and perform kinematics and motion analysis as if a working prototype existed. Despite such strengths of virtual prototyping, there is one inherent weakness that such soft prototypes cannot be tested for phenomena that is not anticipated or accounted for in the computer program. As such there is no guarantee that the virtual prototype is really problem free.

1.2.4   Third Phase: Rapid Prototyping

Rapid Prototyping of physical parts, or otherwise known as solid freeform fabrication or desktop manufacturing or layer manufacturing technology, represents the third phase in the evolution of prototyping. The invention of this series of rapid prototyping methodologies is described as a "watershed event" [11] because of the tremendous time savings, especially for complicated models. Though the parts (individual components) are relatively three times as complex as parts made in 1970s, the time required to make such a part now averages only three weeks [9]. Since 1988, more than twenty different rapid prototyping techniques have emerged.

1.3   FUNDAMENTALS OF RAPID PROTOTYPING

Common to all the different techniques of RP is the basic approach they adopt, which can be described as follows:

(1)   A model or component is modeled on a Computer-Aided Design/ Computer-Aided Manufacturing (CAD/CAM) system. The model which represents the physical part to be built must be represented as closed surfaces which unambiguously define an enclosed volume. This means that the data must specify the inside, outside and boundary of the model. This requirement will become redundant if the modeling technique used is solid modeling. This is by virtue of the technique used, as a valid solid model will automatically be an enclosed volume. This requirement ensures that all horizontal cross sections that are essential to RP are closed curves to create the solid object.

(2)   The solid or surface model to be built is next converted into a format dubbed the "STL" (STereoLithography) file format which originates from 3D Systems. The STL file format approximates the surfaces of the model by polygons. Highly curved surfaces must employ many polygons, which means that STL files for curved parts can be very large. However, there are some rapid prototyping systems which also accept IGES (Initial Graphics Exchange Specifications) data, provided it is of the correct "flavor".

(3) A computer program analyzes a STL file that defines the model to be fabricated and "slices" the model into cross sections. The cross sections are systematically recreated through the solidification of either liquids or powders and then combined to form a 3D model. Another possibility is that the cross sections are already thin, solid laminations and these thin laminations are glued together with adhesives to form a 3D model. Other similar methods may also be employed to build the model.

Fundamentally, the development of RP can be seen in four primary areas. The Rapid Prototyping Wheel in Figure 1.3 depicts these four key aspects of Rapid Prototyping. They are: Input, Method, Material and Applications.

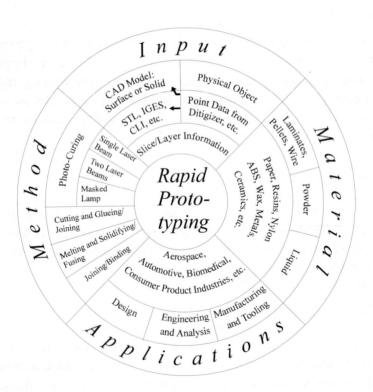

Figure 1.3: The Rapid Prototyping Wheel depicting the four major aspects of RP

1.3.1   Input

Input refers to the electronic information required to describe the physical object with 3D data. There are two possible starting points — a computer model or a physical model. The computer model created by a CAD system can be either a surface model or a solid model. On the other hand, 3D data from the physical model is not at all straightforward. It requires data acquisition through a method known as reverse engineering. In reverse engineering, a wide range of equipment can be used, such as CMM (coordinate measuring machine) or a laser digitizer, to capture data points of the physical model and "reconstruct" it in a CAD system.

1.3.2   Method

While they are currently more than 20 vendors for RP systems, the method employed by each vendor can be generally classified into the following categories: photo-curing, cutting and glueing/joining, melting and solidifying/fusing and joining/binding. Photo-curing can be further divided into categories of single laser beam, double laser beams and masked lamp.

1.3.3   Material

The initial state of material can come in either solid, liquid or powder state. In solid state, it can come in various forms such as pellets, wire or laminates. The current range materials include paper, nylon, wax, resins, metals and ceramics.

1.3.4   Applications

Most of the RP parts are finished or touched up before they are used for their intended applications. Applications can be grouped into (1) Design (2) Engineering, Analysis, and Planning and (3) Tooling and Manufacturing. A wide range of industries can benefit from RP and these include, but are not limited to, aerospace, automotive, biomedical, consumer, electrical and electronics products.

1.4   ADVANTAGES OF RAPID PROTOTYPING

Today's automated, toolless, patternless RP systems can directly produce functional parts in small production quantities. Parts produced in this way usually have an accuracy and surface finish inferior to those made by machining. However, some advanced systems are able to produce near tooling quality parts that are close to or are the final shape. The parts produced, with appropriate post processing, will have material qualities and properties close to the final product. More fundamentally, the time to produce any part — once the design data are available — will be fast, and can be in a matter of hours.

The benefits of RP systems are immense and can be categorized into direct and indirect benefits.

1.4.1   Direct Benefits

The benefits to the company using RP systems are many. One would be the ability to experiment with physical objects of any complexity in a relatively short period of time. It is observed that over the last 25 years, products realized to the market place have increased in complexity in shape and form [9]. For instance, compare the aesthetically beautiful car body of today with that of the 1970s. On a relative complexity scale of 1 to 3 as seen in Figure 1.4, it is noted that from a base of 1 in 1970, this relative complexity index has increased to about 2 in 1980 and close to 3 in the 1990s. More interestingly and ironically, the relative project completion times have not been drastically increased. Initially, from a base of about 4 weeks' project completion time in 1970, it increased to 16 weeks in 1980. However, with the use of CAD/CAM and CNC technologies, project completion time reduces to 8 weeks. Eventually, RP systems allowed the project manager to further cut the completion time to 3 weeks in 1995.

To the individual in the company, the benefits can be varied and have different impacts. It depends on the role in which they play in the

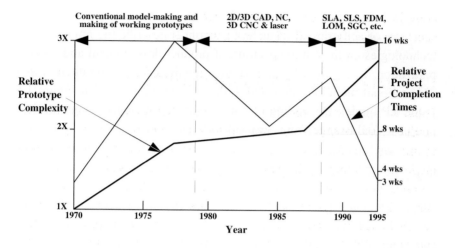

Figure 1.4: Project time and product complexity in 25 years' time frame

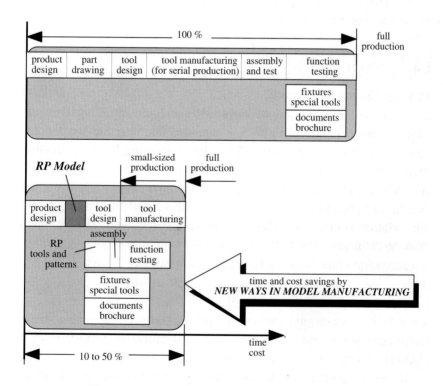

Figure 1.5: Results of the integration of RP technologies

company. The full production of any product encompasses a wide spectrum of activities. Kochan and Chua [12] describe the impact of RP technologies on the entire spectrum of product development and process realization. In Figure 1.5, the activities required for full production in a conventional model are depicted at the top. At the bottom of Figure 1.5 is the RP model. Depending on the size of production, savings on time and cost could range from 50% up to 90%!

1.4.1.1   *Benefits to Product Designers*

The product designers can increase part complexity with little significant effects on lead time and cost. More organic, sculptured shapes for functional or aesthetic reasons can be accommodated. They can optimize part design to meet customer requirements, with little restrictions by manufacturing. In addition, they can reduce parts count by combining features in single-piece parts that are previously made from several because of poor tool accessibility or the need to minimize machining and waste. With fewer parts, time spent on tolerance analysis, selecting fasteners, detailing screw holes and assembly drawings is greatly reduced.

There will also be fewer constraints in the form of parts design without regard to draft angles, parting lines or other such constraints. Parts which cannot easily be set up for machining, or have accurate, large thin walls, or do not use stock shapes to minimize machining and waste can now be designed. They can minimize material and optimize strength/weight ratios without regard to the cost of machining. Finally, they can minimize time-consuming discussions and evaluations of manufacturing possibilities.

1.4.1.2   *Benefits to the Tooling and Manufacturing Engineer*

The main savings are in costs. The manufacturing engineer can minimize design, manufacturing and verification of tooling. He can realize profits earlier on new products, since fixed costs are lower. He can also reduce parts count and, therefore, assembly, purchasing and inventory expenses.

The manufacturer can reduce the labor content of manufacturing, since part-specific setting up and programming are eliminated, machining/casting labor is reduced, and inspection and assembly are also consequently reduced as well. Reducing material waste, waste disposal costs, material transportation costs, inventory cost for raw stock and finished parts (making only as many as required, therefore, reducing storage requirements) can contribute to low overheads. Less inventory is scrapped because of design changes or disappointing sales.

In addition, the manufacturer can simplify purchasing since unit price is almost independent of quantity, therefore, only as many as are needed for the short-term need be ordered. Quotations vary little among supplies, since fabrication is automatic and standardized. One can purchase one general purpose machine rather than many special purpose machines and therefore, reduce capital equipment and maintenance expenses, need fewer specialized operators and less training. A smaller production facility will also result in less effort in scheduling production. Furthermore, one can reduce the inspection reject rate since the number of tight tolerances required when parts must mate can be reduced. One can avoid design misinterpretations (instead, "what you design is what you get"), quickly change design dimensions to deal with tighter tolerances and achieve higher part repeatability, since tool wear is eliminated. Lastly, one can reduce spare parts inventories (produce spare on demand, even for obsolete products).

1.4.2   Indirect Benefits

Outside the design and production departments, indirect benefits can also be derived. Marketing as well as the customers will also benefit from the utilization of RP technologies.

1.4.2.1   *Benefits to Marketing*

To the market, it presents new capabilities and opportunities. It can greatly reduce time-to-market, resulting in (1) reduced risk as there is no need to project customer needs and market dynamics several years into the future, (2) products which fit customer needs much more closely, (3) products offering the price/performance of the latest technology, (4) new products being test-marketed economically.

Marketing can also change production capacity according to market demand, possibly in real time and with little impact on manufacturing. One can increase the diversity of product offerings and pursue market niches currently too small to justify due to tooling cost (including custom and semi-custom production). One can easily expand distribution and quickly enter foreign markets.

1.4.2.2   *Benefits to the Consumer*

The consumer can buy products which meet more closely individual needs and wants. Firstly, there is a much wider diversity of offerings to choose from. Secondly, one can buy (and even contribute to the design of) affordable products built-to-order. Furthermore, the consumer can buy products at lower prices, since the manufacturers' savings will ultimately be passed on.

1.5   COMMONLY USED TERMS

The number of terms used by the engineering communities around the world is alarmingly large. Perhaps, this is due to the newness of the technology. It certainly does not help as already there are so many buzz words used today. Worldwide, the most commonly used term is Rapid Prototyping. The term is apt as the key benefit of RP is its *rapid* creation of a physical model. However, prototyping is slowly growing to include other areas. Soon, *Rapid Prototyping, Tooling and Manufacturing* (RPTM) should be used to include the utilization of the prototype as a master pattern for tooling and manufacturing.

Some of the less commonly used terms include *Direct CAD Manufacturing, Desktop Manufacturing* and *Instant Manufacturing*. The rationale behind these terms are also speed and ease, though not exactly direct or instant! *CAD Oriented Manufacturing* is another term and provides an insight into the issue of orientation, often a key factor influencing the output of a prototype made by RP methods like SLA.

Another group of terms emphasizes on the unique characteristic of RP — layer by layer addition as opposed to traditional manufacturing methods such as machining which is material removal from a block. This group includes *Layer Manufacturing, Material Deposit Manufacturing, Material Addition Manufacturing* and *Material Incress Manufacturing*.

There is yet another group which chooses to focus on the words "solid" and "freeform" — *Solid Freeform Manufacturing* and *Solid Freeform Fabrication. Solid* is used because while the initial state may be liquid, powder, individual pellets or laminates, the end result is a solid, 3D object, while *freeform* stresses on the ability of RP to build complex shapes with little constraint on its form.

1.6   CLASSIFICATION OF RAPID PROTOTYPING SYSTEMS

While there are many ways in which one can classify the numerous RP systems in the market, one of the better ways is to classify RP systems broadly by the initial form of its material, i.e. the material that the prototype or part is built with. In this manner, all RP systems can be easily categorized into (1) liquid-based (2) solid-based and (3) powder-based.

1.6.1   Liquid-Based

Liquid-based RP systems have the initial form of its material in liquid state. Through a process commonly known as curing, the liquid is converted into the solid state. The following RP systems fall into this category:

(1)   3D Systems' Stereolithography Apparatus (SLA)
(2)   Cubital's Solid Ground Curing (SGC)
(3)   Sony's Solid Creation System (SCS)
(4)   CMET's Solid Object Ultraviolet-Laser Printer (SOUP)
(5)   Autostrade's E-Darts
(6)   Teijin Seiki's Soliform System

(7)   Meiko's Rapid Prototyping System for the Jewelry Industry
(8)   Denken's SLP
(9)   Mitsui's COLAMM
(10) Fockele & Schwarze's LMS
(11) Light Sculpting
(12) Aaroflex
(13) Rapid Freeze
(14) Two Laser Beams
(15) Microfabrication

As is illustrated in the RP Wheel in Figure 1.3, three methods are possible under the *"Photo-curing"* method. The *single laser beam* method is most widely used and include all the above RP systems with the exception of (2), (11), (13) and (14). Cubital (2) and Light Sculpting (11) use the *masked lamp* method, while the *two laser beam* method is still not commercialized. Rapid Freeze (13) involves the freezing of water droplets and deposit in a manner much like FDM to create the prototype. Each of these RP systems will be described in more detail in Chapter 3.

1.6.2   Solid-Based

Except for powder, solid-based RP systems are meant to encompass all forms of material in the solid state. In this context, the solid form can include the shape in the form of a wire, a roll, laminates and pellets. The following RP systems fall into this definition:

(1)   Cubic Technologies' Laminated Object Manufacturing (LOM)
(2)   Stratasys' Fused Deposition Modeling (FDM)
(3)   Kira Corporation's Paper Lamination Technology (PLT)
(4)   3D Systems' Multi-Jet Modeling System (MJM)
(5)   Solidscape's ModelMaker and PatternMaster
(6)   Beijing Yinhua's Slicing Solid Manufacturing (SSM), Melted Extrusion Modeling (MEM) and Multi-Functional RPM Systems (M-RPM)

(7) CAM-LEM's CL 100

(8) Ennex Corporation's Offset Fabbers

Referring to the RP Wheel in Figure 1.3, two methods are possible for solid-based RP systems. RP systems (1), (3), (4) and (9) belong to the *Cutting and Glueing/Joining* method, while the *Melting and Solidifying/Fusing* method used RP systems (2), (5), (6), (7) and (8). The various RP systems will be described in more detail in Chapter 4.

1.6.3 Powder-Based

In a strict sense, powder is by-and-large in the solid state. However, it is intentionally created as a category outside the solid-based RP systems to mean powder in grain-like form. The following RP systems fall into this definition:

(1) 3D Systems's Selective Laser Sintering (SLS)

(2) EOS's EOSINT Systems

(3) Z Corporation's Three-Dimensional Printing (3DP)

(4) Optomec's Laser Engineered Net Shaping (LENS)

(5) Soligen's Direct Shell Production Casting (DSPC)

(6) Fraunhofer's Multiphase Jet Solidification (MJS)

(7) Acram's Electron Beam Melting (EBM)

(8) Aeromet Corporation's Lasform Technology

(9) Precision Optical Manufacturing's Direct Metal Deposition (DMD™)

(10) Generis' RP Systems (GS)

(11) Therics Inc.'s Theriform Technology

(12) Extrude Hone's Prometal™ 3D Printing Process

All the above RP systems employ the *Joining/Binding* method. The method of joining/binding differs for the above systems in that some employ a laser while others use a binder/glue to achieve the joining effect. Similarly, the above RP systems will be described in more detail in Chapter 5.

REFERENCES

[1] Wheelwright, S.C. and Clark, K.B., *Revolutionizing Product Development: Quantum Leaps in Speed, Efficiency, and Quality*, The Free Press, New York, 1992.

[2] Ulrich, K.T. and Eppinger, S.D., *Product Design and Development*, 2nd edition, McGraw Hill, Boston, 2000.

[3] Hornby, A.S. and Wehmeier, S. (Editor), *Oxford Advanced Learner's Dictionary of Current English*, 6th edition, Oxford University Press, Oxford, 2000.

[4] Koren, Y., *Computer Control of Manufacturing Systems*, McGraw Hill, Singapore, 1983.

[5] Hecht, J., *The Laser Guidebook*, 2nd edition, McGraw Hill, New York, 1992.

[6] Taraman, K., *CAD/CAM: Meeting Today's Productivity Challenge*, Computer and Automated Systems Association of SME, Michigan, 1982.

[7] Chua, C.K., "Three-dimensional rapid prototyping technologies and key development areas," *Computing and Control Engineering Journal* **5**(4) (1994): 200–206.

[8] Chua, C.K., "Solid modeling — A state-of-the-art report," *Manufacturing Equipment News* (September 1987): 33–34.

[9] Metelnick, J., "How today's model/prototype shop helps designers use rapid prototyping to full advantage," *Society of Manufacturing Engineers Technical Paper* (1991): MS91-475.

[10] Lee, G., "Virtual prototyping on personal computers," *Mechanical Engineering* **117**(7) (1995): 70–73.

[11] Kochan, D., "Solid freeform manufacturing — Possibilities and restrictions," *Computers in Industry* **20** (1992): 133–140.

[12] Kochan, D. and Chua, C.K., "State-of-the-art and future trends in advanced rapid prototyping and manufacturing," *International Journal of Information Technology* **1**(2) (1995): 173–184.

PROBLEMS

1. How would you define prototype in the context of modern product development?

2. What are the three aspects of interest in describing a prototype? Describe them clearly.

3. What are the main roles and functions for prototypes? How do you think rapid prototyping satisfies these roles?

4. Describe the historical development of Rapid Prototyping and related technologies.

5. What are the three phases of prototyping? Contrasting these with those of geometric modeling, what similarities can be drawn?

6. Despite the increase in relative complexity of the shape and form of products, project times has been kept relatively shorter. Why?

7. What are the fundamentals of Rapid Prototyping?

8. What is the *Rapid Prototyping Wheel*? Describe its four primary aspects. Is the *Wheel* a static representation of what is Rapid Prototyping today? Why?

9. Describe the advantages of Rapid Prototyping in terms of its beneficiaries such as the product designers, tool designer, manufacturing engineer, marketeers and consumers?

10. Many terms have been used to mean Rapid Prototyping. Discuss three such terms and explain why they have been used in place of Rapid Prototyping.

11. Name three Rapid Prototyping Systems that are liquid-based.

12. How can the liquid form be converted to the solid form as in these liquid-based Rapid Prototyping Systems?

13. In what form of material can Rapid Prototyping Systems be classified as solid-based? Name three such systems.

14. What is the method used in powder-based Rapid Prototyping Systems?

Chapter 2
RAPID PROTOTYPING PROCESS CHAIN

2.1 FUNDAMENTAL AUTOMATED PROCESSES

There are three fundamental fabrication processes [1, 2] as shown in Figure 2.1. They are *Subtractive*, *Additive* and *Formative* processes.

In the subtractive process, one starts with a single block of solid material larger than the final size of the desired object and material is removed until the desired shape is reached.

In contrast, an additive process is the exact reverse in that the end product is much larger than the material when it started. A material is manipulated so that successive portions of it combine to form the desired object.

Lastly, the formative process is one where mechanical forces or restricting forms are applied on a material so as to form it into the desired shape.

There are many examples for each of these fundamental fabrication processes. Subtractive fabrication processes include most forms of machining processes — CNC or otherwise. These include milling, turning, drilling, planning, sawing, grinding, EDM, laser cutting, water-

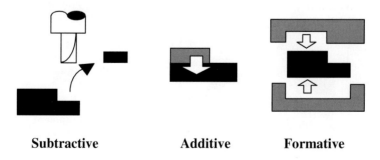

| **Subtractive** | **Additive** | **Formative** |

Figure 2.1: Three types of fundamental fabrication processes

jet cutting and the likes. Most forms of rapid prototyping processes such as Stereolithography and Selective Laser Sintering fall into the additive fabrication processes category. Examples of formative fabrication processes are: Bending, forging, electromagnetic forming and plastic injection molding. These include both bending of sheet materials and molding of molten or curable liquids. The examples given are not exhaustive but indicative of the range of processes.

Hybrid machines combining two or more fabrication processes are also possible. For example, in progressive pressworking, it is common to see a hybrid of subtractive (as in blanking or punching) and formative (as in bending and forming) processes.

2.2   PROCESS CHAIN

As described in Section 1.3, all RP techniques adopt the same basic approach. As such all RP systems generally have a similar sort of process chain. Such a generalized process chain is shown in Figure 2.2 [3]. There are a total of five steps in the chain and these are 3D

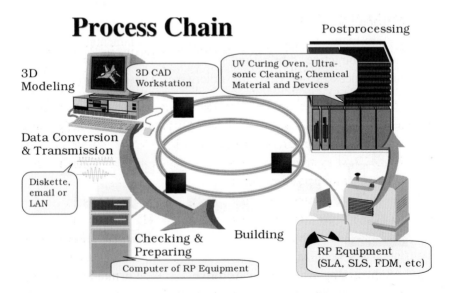

Figure 2.2: Process chain of Rapid Prototyping process

modeling, data conversion and transmission, checking and preparing, building and postprocessing. Depending on the quality of the model and part in Steps 3 and 5 respectively, the process may be iterated until a satisfactory model or part is achieved.

However, like other fabrication processes, process planning is important before the RP commences. In process planning, the steps of the RP process chain are listed. The first step is 3D geometric modeling. In this instance, the requirement would be a workstation and a CAD modeling system. The various factors and parameters which influence the performance of each operation are examined and decided upon. For example, if a SLA is used to build the part, the orientation of the part is an important factor which would, amongst other things, influence the quality of the part and the speed of the process. Needless to say, an operation sheet used in this manner requires proper documentation and guidelines. Good documentation, such as a process logbook, allows future examination and evaluation, and subsequent improvements can be implemented to process planning. The five steps are discussed in the following sections.

2.3   3D MODELING

Advanced 3D CAD modeling is a general prerequisite in RP processes and, usually is the most time-consuming part of the entire process chain. It is most important that such 3D geometric models can be shared by the entire design team for many different purposes, such as interference studies, stress analyses, FEM analysis, detail design and drafting, planning for manufacturing, including NC programming, etc. Many CAD/CAM systems now have a 3D geometrical modeler facility with these special purpose modules.

There are two common misconceptions amongst new users of RP. First, unlike NC programming, RP requires a closed volume of the model, whether the basic elements are surfaces or solids. This confusion arises because new users are usually acquainted with the use of NC programming where a single surface or even a line element can be an NC element. Second, new users also usually assume *what you see is what you get*. These two misconceptions often lead to under-specifying parameters to the RP systems, resulting in poor performance and nonoptimal utilization of the system. Examples of considerations that have to be taken into account include orientation of part, need for supports, difficult-to-build part structure such as thin walls, small slots or holes and overhanging elements. Therefore, RP users have to learn and gain experience from working on the system. The problem is usually more complex than one can imagine because there are many different RP machines which have different requirements and capabilities. For example, while a SLA requires supports, SGC does not, and SGC works most economically if many parts are nested together and processed simultaneously (see Chapter 3, Sections 3.1 and 3.2).

2.4   DATA CONVERSION AND TRANSMISSION

The solid or surface model to be built is next converted into a format dubbed the STL file format. This format originates from 3D Systems which pioneers the **ST**ereo**L**ithography system. The STL file format approximates the surfaces of the model using tiny triangles. Highly curved surfaces must employ many more triangles, which mean that STL files for curved parts can be very large. The STL file format will be discussed in detail in Chapter 6.

Almost, if not all, major CAD/CAM vendors supply the CAD-STL interface. Since 1990, almost all major CAD/CAM vendors have developed and integrated this interface into their systems.

This conversion step is probably the simplest and shortest of the entire process chain. However, for a highly complex model coupled with an extremely low performance workstation or PC, the conversion can take several hours. Otherwise, the conversion to STL file should take only several minutes. Where necessary, supports are also converted to a separate STL file. Supports can alternatively be created or modified in the next step by third party software which allows verification and modifications of models and supports.

The transmission step is also fairly straightforward. The purpose of this step is to transfer the STL files which reside in the workstation to the RP system's computer. It is typical that the workstation and the RP system are situated in different locations. The workstation, being a design tool, is typically located in a design office. The RP system, on the other hand, is a process or production machine, and is usually located on the shopfloor. Data transmission via agreed data formats such as STL or IGES may be carried out through a diskette, email (electronic mail) or LAN (local area network). No validation of the quality of the STL files is carried out at this stage.

2.5   CHECKING AND PREPARING

The computer term, *garbage in garbage out*, is also applicable to RP. Many first time users are frustrated at this step to discover that their STL files are faulty. However, more often than not, it is due to both the errors of CAD models and the nonrobustness of the CAD-STL interface. Unfortunately, today's CAD models — whose quality are dependent on the CAD systems, human operators and postprocesses — are still afflicted with a wide spectrum of problems, including the generation of unwanted shell-punctures (i.e. holes, gaps, cracks, etc.). These problems, if not rectified, will result in the frequent failure of applications downstream. These problems are discussed in detail in the first few sections of Chapter 6.

At present, the CAD model errors are corrected by human operators assisted by specialized software such as MAGICS, a software developed by Materialise, N.V., Belgium [4]. This process of manual repair is very tedious and time consuming especially if one considers the great number of geometric entities (e.g. triangular facets) that are encountered in a CAD model. The types of errors and its possible solutions are discussed in Chapter 6.

Once the STL files are verified to be error-free, the RP system's computer analyzes the STL files that define the model to be fabricated and slices the model into cross-sections. The cross-sections are systematically recreated through the solidification of liquids or binding of powders, or fusing of solids, to form a 3D model.

In a SLA, for example, each output file is sliced into cross-sections, between 0.12 mm (minimum) to 0.50 mm (maximum) in thickness. Generally, the model is sliced into the thinnest layer (approximately 0.12 mm) as they have to be very accurate. The supports can be created using coarser settings. An internal cross hatch structure is generated between the inner and the outer surface boundaries of the part. This serves to hold up the walls and entrap liquid that is later solidified with the presence of UV light.

Preparing building parameters for positioning and stepwise manu-facturing in the light of many available possibilities can be difficult if not accompanied by proper documentation. These possibilities include determination of the geometrical objects, the building orientation, spatial assortments, arrangement with other parts, necessary support structures and slice parameters. They also include the determination of technological parameters such as cure depth, laser power and other physical parameters as in the case of SLA. It means that user-friendly software for ease of use and handling, user support in terms of user manuals, dialogue mode and online graphical aids will be very helpful to users of the RP system.

Many vendors are continually working to improve their systems in this aspect. For example, a software, *Partman Program*, was introduced by 3D Systems [5] to reduce the time spent on setting parameters for

Table 2.1: Parameters used in the SLA process

1.	X–Y shrink
2.	Z shrink
3.	Number of copies
4.	Multi-part spacing
5.	Range manager (add delete, etc.)
6.	Recoating
7.	Slice output scale
8.	Resolution
9.	Layer thickness
10.	X–Y hatch-spacing or 60/120 hatch spacing
11.	Skin fill spacing (X, Y)
12.	Minimum hatch intersecting angle

the SLA process. Before this software is introduced, parameters (such as the location in the 250 mm × 250 mm box and the various cure depths) had to be set manually. This was very tedious for there may be up to 12 parameters to be keyed in. These parameters are shown in Table 2.1.

However, the job is now made simpler with the introduction of default values that can be altered to other specific values. These values can be easily retrieved for use in other models. This software also allows the user to orientate and move the model such that the whole model is in the positive axis' region (the SLA uses only positive numbers for calculations). Thus the original CAD design model can also be in "negative" regions when converting to STL format.

2.6   BUILDING

For most RP systems, this step is fully automated. Thus, it is usual for operators to leave the machine on to build a part overnight. The building process may take up to several hours to build depending on the size and number of parts required. The number of identical parts that can be built is subject to the overall build size constrained by the build volume of the RP system.

2.7   POSTPROCESSING

The final task in the process chain is the postprocessing task. At this stage, generally some manual operations are necessary. As a result, the danger of damaging a part is particularly high. Therefore, the operator for this last process step has a high responsibility for the successful process realization. The necessary postprocessing tasks for some major RP systems are shown in Table 2.2.

The cleaning task refers to the removal of excess parts which may have remained on the part. Thus, for SLA parts, this refers to excess resin residing in entrapped portion such as a blind hole of a part, as well as the removal of supports. Similarly, for SLS parts, the excess powder has to be removed. Likewise for LOM, pieces of excess wood-like blocks of paper which acted as supports have to be removed.

Table 2.2: Essential postprocessing tasks for different RP processes
(✓ = is required; ✗ = not required)

Rapid Prototyping Technologies				
Postprocessing Tasks	SLS[1]	SLA[2]	FDM[3]	LOM[4]
1. Cleaning	✓	✓	✗	✓
2. Postcuring	✗	✓	✗	✗
3. Finishing	✓	✓	✓	✓

[1]SLS — Selective Laser Sintering
[2]SLA — Stereolithography Apparatus
[3]FDM — Fused Deposition Modeling
[4]LOM — Layered Object Manufacturing

As shown in Table 2.2, the SLA procedures require the highest number of postprocessing tasks. More importantly, for safety reason, specific recommendations for postprocessing tasks have to be prepared, especially for cleaning of SLA parts. It was reported that accuracy is related to the post-treatment process [6]. Specifically, Ref. 6 refers to the swelling of SLA-built parts with the use of cleaning solvents. Parts are typically cleaned with solvent to remove unreacted photosensitive resin. Depending upon the "build style" and the extent of crosslinking in the resin, the part can be distorted during the cleaning process. This effect was particularly pronounced with the more open "build styles" and aggressive solvents. With the "build styles" approaching a solid fill and more solvent-resistant materials, damage with the cleaning solvent can be minimized. With newer cleaning solvents, like TPM (tripropylene glycol monomethyl ether) introduced by 3D Systems, part damage due to the cleaning solvent can be reduced or even eliminated [6].

For reasons which will be discussed in Chapter 3, SLA parts are built with pockets of liquid embedded within the part. Therefore, postcuring is required. All other nonliquid RP methods do not undergo this task.

Finishing refers to secondary processes such as sanding and painting used primarily to improve the surface finish or aesthetic appearance of the part. It also includes additional machining processes such as drilling, tapping and milling to add necessary features to the parts.

REFERENCES

[1] Burns, M., Research Notes. *Rapid Prototyping Report* **4**(3) (1994): 3–6.

[2] Burns, M., *Automated Fabrication*, PTR Prentice Hall, New Jersey, 1993.

[3] Kochan, D. and Chua, C.K., "Solid freeform manufacturing — Assessments and improvements at the entire process chain," *Proceedings of the International Dedicated Conference on Rapid Prototyping for the Automotive Industries*, ISATA94, Aachen, Germany, 31 Oct to 4 Nov 94.

[4] Materialise, N.V., Magics 3.01 Materialise User Manual. Materalise Software Department, Kapeldreef 60, B-3001 Heverlee, Belgium, 1994.

[5] 3D Systems, *SLA User Reference Manual*.

[6] Peiffer, R.W., "The laser stereolithography process — Photosensitive materials and accuracy," *Proceedings of the First International User Congress on Solid Freeform Manufacturing*, Germany, 28–30 Oct 1993.

PROBLEMS

1. What are the three types of automated fabricators? Describe them and give two examples each.

2. Each one of the following manufacturing processes/methods in Table 2.3 belongs to one of the three basic types of fabricators. Tick [✓] under the column if you think it belongs to that category. If you think that it is a hybrid machine, you may tick [✓] in more than one category.

Table 2.3: Problem in fundamental fabrication processes

S/No	Manufacturing Process	Subtractive	Additive	Formative
1.	Pressworking			
2.	*SLS			
3.	Plastic Injection Molding			
4.	CNC Nibbling			
5.	*CNC CMM			
6.	*LOM			

* For a list of abbreviations used, please refer to the front part of the book.

3. Describe the five steps involved in a general RP process chain. Which steps do you think are likely to be iterated?

4. After 3D geometric modeling, a user can either make a part through NC programming or through rapid prototyping. What are the basic differences between NC programming and RP in terms of the CAD model?

5. STL files are problematic. Is this a fair statement to make? Discuss.

6. Preparing for building appears to be fairly sophisticated. In the case of a SLA, what are some of the considerations and parameters involved?

7. Distinguish cleaning, postcuring and finishing which are the various tasks of postprocessing. Name two RP processes that do not require postcuring and one that does not require cleaning.

8. Which step in the entire process chain is, in your opinion, the shortest? Most tedious? Most automated? Support your choice.

Chapter 3
LIQUID-BASED RAPID PROTOTYPING SYSTEMS

Most liquid-based rapid prototyping systems build parts in a vat of photo-curable liquid resin, an organic resin that cures or solidifies under the effect of exposure to laser radiation, usually in the UV range. The laser cures the resin near the surface, forming a hardened layer. When a layer of the part is formed, it is lowered by an elevation control system to allow the next layer of resin to be similarly formed over it. This continues until the entire part is completed. The vat can then be drained and the part removed for further processing, if necessary. There are variations to this technique by the various vendors and they are dependent on the type of light or laser, method of scanning or exposure, type of liquid resin, type of elevation and optical system used.

3.1 3D SYSTEMS' STEREOLITHOGRAPHY APPARATUS (SLA)

3.1.1 Company

3D Systems was founded in 1986 by inventor Charles W. Hull and entrepreneur Raymond S. Freed. Amongst all the commercial RP systems, the Stereolithography Apparatus, or SLA® as it is commonly called, is the pioneer with its first commercial system marketed in 1988. It has been awarded more than 40 United States patents and 20 international patents, with additional patents filed or pending inter-nationally. 3D Systems Inc. is currently headquartered in 26801 Avenue Hall, Valencia, CA 91355, USA.

3.1.2   Products

3.1.2.1   *Models and Specifications*

3D Systems produces a wide range of machines to cater to various part sizes and throughput. There are several models available, including those in the series of SLA 250/30A, SLA 250/50, SLA-250/50HR, SLA 3500, SLA 5000, SLA 7000 and Viper si2. The SLA 250/30A is an economical and versatile SLA starter system that uses a Helium Cadmium (He–Cd) laser. The SLA 250/50 is a supercharged system with a higher powered laser, interchangeable vats and Zephyr recoater system, whereas the SLA 250/50HR adds a special feature of a small spot laser for high-resolution application. All SLA 250 type systems have a maximum build envelope of 250 × 250 × 250 mm and use a He–Cd laser. For bigger build envelopes, the SLA 3500, SLA 5000 and SLA 7000 are available. These three machines use a different laser from the SLA 250 (solid-state Nd:YVO$_4$). The SLA 7000 (see Figure 3.1) is the top of the series. It can build parts up to four times faster than the SLA 5000 with the capacity of building thinner layers (minimum layer thickness 0.025 mm) for finer surface finish. Its faster speed is largely due to its dual spot laser's

Figure 3.1: 3D Systems' SLA 7000 (Courtesy 3D Systems)

ability. This means that a smaller beam spot is used for the border for accuracy, whereas the bigger beam spot is used for internal cross-hatching for speed. 3D Systems' new Viper si2 SLA system is their first solid imaging system to combine standard and high-resolution part building in the same system. The Viper si2 system lets you choose between standard resolution, for the best balance of build speed and part resolution, and high resolution (HR mode) for ultra-detailed small parts and features. All these are made possible by a carefully integrated digital signal processor (DSP) controlled high speed scanning system with a single, solid-state laser that delivers a constant 100 mW of available power throughout its 7500-hour warranty life. The Viper si2 system builds parts with a smooth surface finish, excellent optical clarity, high accuracy, and thin, straight vertical walls. It is ideal for a myriad of solid imaging applications, from rapid modeling and prototyping to injection molding and investment casting. Specifications of these machines are summarized in Tables 3.1(a) and 3.1(b).

The Zephyr™ system which was introduced in 1996 as a product enhancement to its popular SLA-250 [1] is now used in all the SLA systems with the exception of SLA 250/30A. The Zephyr™ system eliminates the need for the traditional "deep dip" in which a part is dunked into the resin vat after each layer and then raised to within one layer's depth of the top of the vat. With the deep dip, a wiper blade sweeps across the surface of the vat to remove excess resin before scanning the next layer. The Zephyr™ system has a vacuum blade that picks up resin from the side of the vat and applies a thin layer of resin as it sweeps across the part. This speeds up the build process by reducing time required between layers and greatly reduces problems involved when building parts with trapped volumes.

All these machines use one-component, photo-curable liquid resins as the material for building. There are several grades of resins available and the usage is dependent on the laser on the machine and the mechanical requirement of the part. Specific details of the correct type of resins to be used are available from the manufacturer. The other main consumable used on these machines is the cleaning solvent which is required to clean the part of any residual resin after the building of the part is completed on the machine.

Table 3.1(a): Summary specifications of SLA-250 machines (Source from 3D Systems)

Model	SLA 250/30A	SLA 250/50	SLA 250/50HR
SYSTEM CHARACTERISTICS			
	SmartStart. An economical and versatile SLA starter system.	A supercharged system with higher powered laser, interchangeable vats, and Zephyr recoating system.	A specialty system with small spot laser for high-resolution applications.
VAT CAPACITY			
Maximum Build Envelope	$250 \times 250 \times 250$ mm^3 $10 \times 10 \times 10$ in^3	$250 \times 250 \times 250$ mm^3 $10 \times 10 \times 10$ in^3	$250 \times 250 \times 250$ mm^3 $10 \times 10 \times 10$ in^3
VOLUME			
L (U.S. gal)	29.4 (7.8)	32.2 (8.5)	32.2 (8.5)
LASER			
Type	Helium Cadmium (He–Cd)	Helium Cadmium (He–Cd)	Helium Cadmium (He–Cd)
Wavelength	325 nm	325 nm	325 nm
Power at Vat @ hrs	@ 2000/hrs 12 mW	@ 2000/hrs 25 mW	@ 2000/hrs 6 mW
Warranty	2000 hrs	2000 hrs	2000 hrs
OPTICAL & SCANNING			
Dual Spot	No	No	No
Beam Diameter; Border @ $1/e^2$	0.24 +/- 0.04 mm (0.0095 +/- 0.0015 in)	0.24 +/- 0.04 mm (0.0095 +/- 0.0015 in)	0.07 +/- 0.01 mm (0.003 +/- 0.0005 in)
Beam Diameter; Hatch @ $1/e^2$	0.24 +/- 0.04 mm (0.0095 +/- 0.0015 in)	0.24 +/- 0.04 mm (0.0095 +/- 0.0015 in)	0.07 +/- 0.01 mm (0.003 +/- 0.0005 in)
RECOATING SYSTEM			
	Doctor	Zephyr	Zephyr

Table 3.1(a): (*Continued*)

Model	SLA 250/30A	SLA 250/50	SLA 250/50HR
FEATURES			
Interchangeable Vat	Available Option	Yes	Yes
SmartSweep	No	No	No
Auto Resin Refill	No	No	No
SOFTWARE			
3D Lghtyear / Windows NT	With Build-station 3.8.4	With Build-station 3.8.4	With Build-station 3.8.4
Buildstation O/S	MS DOS	MS DOS	MS DOS
RESINS			
General Purpose	SL 5149, SL 5170, SL 5220	SL 5149, SL 5170, SL 5220	SL 5149, SL 5170, SL 5220
Durable	N/A	N/A	N/A
High Temperature	SL 5210	SL 5210	SL 5210
WARRANTY			
	1 yr from installation date	1 yr from installation date	1 yr from installation date

Table 3.1(b): Summary specifications of the rest of the SLA machines (Source from 3D Systems)

Model	SLA 3500	SLA 5000	SLA 7000	Viper si2
		SYSTEM CHARACTERISTICS		
	A mid-sized system up to 2.5 times faster than SLA 250 with productivity enhancements like auto resin refill and SmartSweep.	A large-frame system with three times the build volume of SLA 3500.	A supercharged large-frame system two times faster than SLA 5000 with the capability of building thinner layers for finer surface finish.	A dual-resolution, constant power, longer-life laser.
		VAT CAPACITY		
Maximum Build Envelope	$350 \times 350 \times 400$ mm^3 $13.8 \times 13.8 \times 15.7$ in^3	$508 \times 508 \times 584$ mm^3 $20 \times 20 \times 23$ in^3	$508 \times 508 \times 600$ mm^3 $20 \times 20 \times 23.6$ in^3	$250 \times 250 \times 250$ mm^3 $10 \times 10 \times 10$ in^3
		VOLUME		
L (U.S. gal)	99.3 (25.6)	253.6 (67)		32.2 (8.5l)
		LASER		
Type		Solid-State (Nd:YVO$_4$)		
Wavelength		354.7 nm		
Power at Vat @ hrs	@ 5000/hrs 160 mW	@ 5000/hrs 216 mW	@ 5000/hrs 800 mW	@ 7500/hrs 100 mW
Warranty		5000 hrs		7500 hrs
		OPTICAL & SCANNING		
Dual Spot	No		Yes	
Beam Diameter; Border @ $1/e^2$	0.25 +/- 0.025 mm (0.010 +/- 0.001 in)	0.25 +/- 0.025 mm (0.010 +/- 0.001 in)		0.25 +/- 0.025 mm (0.010 +/- 0.001 in)
Beam Diameter; Hatch @ $1/e^2$	0.25 +/- 0.025 mm (0.010 +/- 0.001 in)		0.7615 +/- 0.0765 mm (0.03 +/- 0.003 in)	0.075 +/- 0.015 mm (0.0030 +/- 0.0006 in)
		RECOATING SYSTEM		
		Zephyr		

Table 3.1(b): (*Continued*)

Model	SLA 3500	SLA 5000	SLA 7000	Viper si2
FEATURES				
Interchangeable Vat		Yes		
SmartSweep		Yes		No
Auto Resin Refill		Yes		No
SOFTWARE				
3D Lghtyear / Windows NT		Buildstation 5.1		Buildstation 5.2
Buildstation O/S		Windows NT 3.5.1		Windows NT 4.0
RESINS				
General Purpose	SL 5190, SL 5510	SL 5195, SL 5510	SL 7510	SL 5510
Durable	SL 5520		SL 7540	
High Temperature		SL 5530 HT		N/A
WARRANTY				
		1 yr from installation date		

3.1.2.2 *Advantages and Disadvantages*

The main advantages of using SLA are:

(1) *Round the clock operation.* The SLA can be used continuously and unattended round the clock.
(2) *Good user support.* The computerized process serves as a good user support.
(3) *Build volumes.* The different SLA machines have build volumes ranging from small to large to suit the needs of different users.
(4) *Good accuracy.* The SLA has good accuracy and can thus be used for many application areas.
(5) *Surface finish.* The SLA can obtain one of the best surface finishes amongst RP technologies.
(6) *Wide range of materials.* There is a wide range of materials, from general-purpose materials to specialty materials for specific applications.

The main disadvantages of using SLA are:

(1) *Requires support structures.* Structures that have overhangs and undercuts must have supports that are designed and fabricated together with the main structure.
(2) *Requires post-processing.* Post-processing includes removal of supports and other unwanted materials, which is tedious, time-consuming and can damage the model.
(3) *Requires post-curing.* Post-curing may be needed to cure the object completely and ensure the integrity of the structure.

3.1.3 Process

3D Systems' stereolithography process creates three-dimensional plastic objects directly from CAD data. The process begins with the vat filled with the photo-curable liquid resin and the elevator table set just below the surface of the liquid resin (see Figure 3.2). The operator loads a three-dimensional CAD solid model file into the system. Supports are designed to stabilize the part during building. The translator converts the CAD data into a STL file. The control unit slices the model and

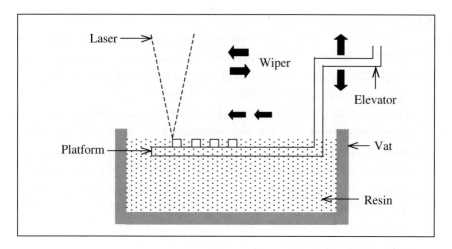

Figure 3.2: Schematic of SLA process

support into a series of cross sections from 0.025 to 0.5 mm (0.001 to 0.020 in) thick. The computer-controlled optical scanning system then directs and focuses the laser beam so that it solidifies a two-dimensional cross-section corresponding to the slice on the surface of the photo-curable liquid resin to a depth greater than one layer thickness. The elevator table then drops enough to cover the solid polymer with another layer of the liquid resin. A leveling wiper or vacuum blade (for Zephyr™ recoating system) moves across the surfaces to recoat the next layer of resin on the surface. The laser then draws the next layer. This process continues building the part from bottom up, until the system completes the part. The part is then raised out of the vat and cleaned of excess polymer.

The main components of the SLA system are a control computer, a control panel, a laser, an optical system and a process chamber. The workstation software used by the SLA system, known as 3D Lightyear exploits the full power of the Windows NT operating system, and delivers far richer functionality than the UNIX-based Maestro software. Maestro includes the following software modules [2]:

(1) *3dverify™ Module.* This module can be accessed to confirm the integrity and/or provide limited repair to stereolithography (STL)

files before part building without having to return to the original CAD software. Gaps between triangles, overlapping or redundant triangles and incorrect normal directions are some examples of the flaws that can be identified and corrected (see Chapter 6).

(2) *View*™ *Module*. This module can display the STL files and slice file (SLI) in graphical form. The viewing function is used for visual inspection and for the orientation of these files so as to achieve optimal building.

(3) *MERGE Module*. By using MERGE, several SLI files can be merged into a group which can be used together in future process.

(4) *Vista*™ *Module*. This module is a powerful software tool that automatically generates support structures for the part files. Support structures are an integral part to successful part building, as they help to anchor parts to the platform when the part is free floating or there is an overhang.

(5) *Part Manager*™ *Module*. This software module is the first stage of preparing a part for building. It utilizes a spreadsheet format into which the STL file is loaded and set-up with the appropriate build and recoat style parameters.

(6) *Slice*™ *Module*. This is the second stage of preparing a part for building. It converts the spreadsheet information from the *Part Manager*™ module to a model of three-dimensional cross sections or layers.

(7) *Converge*™ *Module*. This is the third and last stage of preparing a part for building. This is the module which creates the final build files used by the SLA.

3.1.4   Principle

The SLA process is based fundamentally on the following principles [3]:

(1) Parts are built from a photo-curable liquid resin that cures when exposed to a laser beam (basically, undergoing the photopolymerization process) which scans across the surface of the resin.

(2) The building is done layer by layer, each layer being scanned by the optical scanning system and controlled by an elevation mechanism which lowers at the completion of each layer.

These two principles will be briefly discussed in this section to lay the foundation to the understanding of RP processes. They are mostly applicable to the liquid-based RP systems described in this chapter. This first principle deals mostly with photo-curable liquid resins, which are essentially photopolymers and the photopolymerization process. The second principle deals mainly with CAD data, the laser, and the control of the optical scanning system as well as the elevation mechanism.

3.1.4.1 *Photopolymers*

There are many types of liquid photopolymers that can be solidified by exposure to electro-magnetic radiation, including wavelengths in the gamma rays, X-rays, UV and visible range, or electron-beam (EB) [4, 5]. The vast majority of photopolymers used in the commercial RP systems, including 3D Systems' SLA machines are curable in the UV range. UV-curable photopolymers are resins which are formulated from photoinitiators and reactive liquid monomers. There are a large variety of them and some may contain fillers and other chemical modifiers to meet specified chemical and mechanical requirements [6]. The process through which photopolymers are cured is referred to as the photo-polymerization process.

3.1.4.2 *Photopolymerization*

Loosely defined, polymerization is the process of linking small mole-cules (known as monomers) into chain-like larger molecules (known as polymers). When the chain-like polymers are linked further to one another, a cross-linked polymer is said to be formed. Photopolymeri-zation is polymerization initiated by a photochemical process whereby the starting point is usually the induction of energy from the radiation source [7].

Polymerization of photopolymers is normally an energetically favorable or exothermic reaction. However, in most cases, the formu-lation of a photopolymer can be stabilized to remain unreacted at ambient temperature. A catalyst is required for polymerization to take place at a reasonable rate. This catalyst is usually a free radical which may be generated either thermally or photochemically. The source of a photochemically generated radical is a photoinitiator, which reacts with an actinic photon to produce the radicals that catalyze the polymerization process.

The free-radical photopolymerization process is schematically presented in Figure 3.3 [8]. Photoinitiator molecules, P_i, which are mixed with the monomers, M, are exposed to a UV source of actinic photons, with energy of hv. The photoinitiators absorb some of the photons and are in an excited state. Some of these are converted into reactive initiator molecules, P•, after undergoing several complex chemical energy transformation steps. These molecules then react with a monomer molecule to form a polymerization initiating molecule, PM•. This is the chain initiation step. Once activated, additional monomer molecules go on to react in the chain propagation step, forming longer molecules, PMMM• until a chain inhibition process terminates the polymerization reaction. The longer the reaction is sustained, the higher will be the molecular weight of the resulting polymer. Also, if the monomer molecules have three or more reactive chemical groups, the resulting polymer will be cross-linked, and this will generate an insoluble continuous network of molecules.

During polymerization, it is important that the polymers are sufficiently cross-linked so that the polymerized molecules do not re-dissolve back into the liquid monomers. The photopolymerized molecules must also possess sufficient strength to remain structurally sound while the cured resin is subjected to various forces during recoating.

While free-radical photopolymerization is well-established and yields polymers that are acrylate-based, there is another newer "chemistry" known as cationic photopolymerization [9]. It relies on cationic initiators, usually iodinium or sulfonium salts, to start polymerization. Commercially available cationic monomers include epoxies, the most versatile of cationally polymerizable monomers, and vinylethers. Cationic resins are attractive as prototype materials as they have better physical and mechanical properties. However the process may require higher exposure time or a higher power laser.

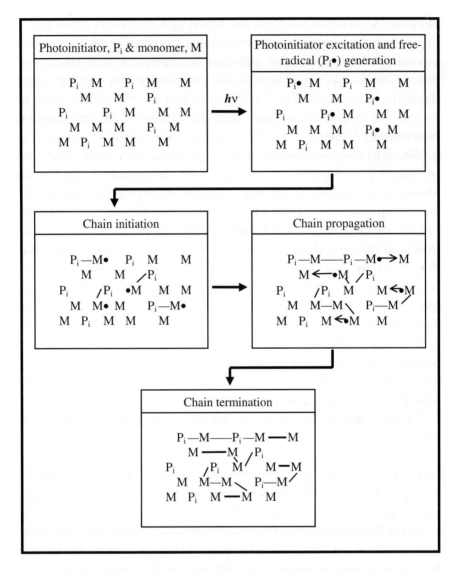

Figure 3.3: Schematic for a simplified free-radical photopolymerization

3.1.4.3 *Layering Technology, Laser and Laser Scanning*

Almost all RP systems use layering technology in the creation of prototype parts (see Chapter 2). The basic principle is the availability of computer software to slice a CAD model into layers and reproduce it in an "output" device like a laser scanning system. The layer thickness is controlled by a precision elevation mechanism. It will correspond directly to the slice thickness of the computer model and the cured thickness of the resin. The limiting aspect of the RP system tends to be the curing thickness rather than the resolution of the elevation mechanism.

The important component of the building process is the laser and its optical scanning system. The key to the strength of the SLA is its ability to rapidly direct focused radiation of appropriate power and wavelength onto the surface of the liquid photopolymer resin, forming patterns of solidified photopolymer according to the cross-sectional data generated by the computer [10]. In the SLA, a laser beam with a specified power and wavelength is sent through a beam expanding telescope to fill the optical aperture of a pair of cross axis, galvanometer driven, beam scanning mirrors. These form the optical scanning system of the SLA. The beam comes to a focus on the surface of a liquid photopolymer, curing a predetermined depth of the resin after a controlled time of exposure (inversely proportional to the laser scanning speed).

The solidification of the liquid resin depends on the energy per unit area (or "exposure") deposited during the motion of the focused spot on the surface of the photopolymer. There is a threshold exposure that must be exceeded for the photopolymer to solidify.

To maintain accuracy and consistency during part building using the SLA, the cure depth and the cured line width must be controlled. As such, accurate exposure and focused spot size become essential.

Parameters which influence performance and functionality of the parts are the physical and chemical properties of the resin, the speed and resolution of the optical scanning system, the power, wavelength and type of the laser used, the spot size of the laser, the recoating system, and the post-curing process.

3.1.5   Applications

The SLA technology provides manufacturers with cost justifiable methods for reducing time to market, lowering product development costs, gaining greater control of their design process and improving product design. The range of applications include:

(1) Models for conceptualization, packaging and presentation.
(2) Prototypes for design, analysis, verification and functional testing.
(3) Parts for prototype tooling and low volume production tooling.
(4) Patterns for investment casting, sand casting and molding.
(5) Tools for fixture and tooling design, and production tooling.

Software developed to support these applications include QuickCast™, a software tool which is used in the investment casting industry. QuickCast enables highly accurate resin patterns that are specifically used as an expendable pattern to form a ceramic mold to be created. The expendable pattern is subsequently burnt out. The standard process uses an expendable wax pattern which must be cast in a tool. QuickCast eliminates the need for the tooling used to make the expendable patterns. QuickCast produces parts which have a hard thin outer shell and contain a honeycomb like structure inside, allowing the pattern to collapse when heated instead of expanding, which would crack the shell.

3.1.6 Examples

3.1.6.1 *Ford Uses Stereolithography to Cast Prototype Tooling* [11]

With a single project triumph, Ford has begun a new era of "rapid manufacturing" — applying QuickCast technology to the development of both prototype and, ultimately, production tooling. Such innovation demonstrates the potential to save manufacturing industries millions in tooling costs.

Ingenuity was Ford's most critical ingredient. When production units of a rear wiper motor cover for the 1994 Explorer were needed for testing, several tooling alternatives were explored. Traditional methods would have provided the tool in three months. Ford used QuickCast in its first application of rapid tooling by investment casting stereolithography mold halves to create the hard tool.

They first built a SL model of the cover, fit it over the wiper motor to verify design integrity, and found a clearance problem. The plastic part was modified by hand, the fit was re-verified, and the CAD model was adjusted. Pro/MOLDESIGN software was then used to create "negative" mold halves from the same CAD data. Shrink factors were applied to compensate for the SL resin, A2 steel, and polypropylene end product material.

The SLA 250-generated QuickCast patterns resulted in a core and cavity pair investment cast in A2 steel. Knowledge of the cast metal's characteristics facilitated changes in a second set of production tooling, e.g., inclusion of ejector holes and addition of cooling lines. The turnaround time for the second set of tooling was only four weeks, and the cost for "QuickCast Tooling" was only $5000 per tool set, compared to the $33 000 quoted for machining a single tool. Ford was able to start durability and water flow testing 18 months ahead of schedule, with costs reduced by 45% and time savings of more than 40% achieved.

3.1.6.2   *Black & Decker Saves a Year by Using Stereolithography to Prototype their Improved Shearer/Shrub Trimmer Power Tool*

Designers at Black & Decker had only 100 days to transform an idea for an improved shearer/shrub trimmer power tool into an attractive functioning prototype that could be introduced at an important trade show. The new power tool, which would expand the company's VersaPak outdoor power tool product line, had to undergo design, proofing, building, assembly and testing of at least 30 copies of the product in that time frame. Failing to deliver functioning prototypes would put Black & Decker back one year in introducing the product to the market place.

The new concept would offer a more ergonomically designed shearer/shrub trimmer. This model would also be more streamlined in appearance. The new design was also to incorporate, in the single body of the trimmer, an easy-access battery pop-out.

Black & Decker turned to Mass Engineered Design Inc., of Toronto, Ontario, where engineers used stereolithography (SL). SL dealt quickly with the complexity of the design features and was the only option that Mass had, to design and build 30 prototypes of the product in the 100-day time frame. A number of CAD and SL iterations were created to identify the best location for the patented battery pop-out design. The SL masters also produced prototypes on which to perform failure analysis of the injection-molding process. The first 30 cast urethane parts received paint and logo labels, and were assembled together with mechanical components to make them functioning power tools.

SL gives Black & Decker the ability to:

(1) Present new products before actual production begins.
(2) Meet delivery times without compromising quality and functionality.
(3) Iterate and verify the designs before committing to the product.
(4) Quickly test new product ideas with functioning prototypes.
(5) Evaluate individual components for fit during product assembly.

Mass Engineered Design Inc., working from product feature specifications, preliminary drawings and a hand-crafted model provided by Black & Decker, began by building a 3D CAD model. X-ray tomography digitized the surface geometry of the hand-crafted model to create a CAD surface model, which was combined with product specifications to create a number of 3D CAD models.

A design team that included toolmakers, molders and material suppliers, critiqued all CAD models. Approved design features were incorporated into the primary CAD file.

A SLA 500 built a master of product components from the approved CAD file. From these masters, Mass technicians created reusable rubber (RTV) molds, from which Mass ultimately vacuum-molded enough urethane plastic components to build 110 prototypes.

The initial 30 prototypes were painted, labeled with logos and assembled together with mechanical components. The now-functioning prototypes were tested, packaged and shipped before the 100-day deadline and in time for the trade show.

The attractive and functioning prototype products (see Figure 3.4) were a success at the trade show, garnering a number of sales orders.

Figure 3.4: Improved shearer/shrub trimmer prototype (Courtesy 3D Systems)

Black & Decker could move the now successful product into full production.

3.1.6.3   *Bose Saves Five Weeks Using Stereolithography Over Traditional Hard Tooling*

Bose Corporation, world leader in audio components and systems located in Framingham, Massachusetts, needed appearance parts for two different car speaker grills, one being 101 mm (4 in) in diameter and the other 76 mm (3 in) in diameter, for actual road testing on an Oldsmobile Aurora. Needing more than an SL prototype, Bose approached Drew Santin, owner of Santin Engineering, and requested a short run of production parts in end-use material which included uniform texture and multiple colors to match the car interior color options. The catch? Parts needed to be delivered in exactly four weeks. Santin determined that traditional hard tooling would take nine weeks to complete due to each part's contoured design, compound curvature parting line, and complex detail. Because of its ability to produce a textured surface finish, its greater temperature control of the tool, and its durability to guarantee a run of at least 500 parts, Santin turned to the Shaw process of rapid tooling for successful delivery of these time-sensitive parts.

After receiving the file transfer from Bose, Sai
building the master part pattern on an SLA 250 using
5170 resin. After the SL part was completed, rubber
made of the part using flexible silicone rubber, cre____
cavity mold. Santin then removed the SL part and sent the rubber mold
impressions to a local foundry where ceramic material was poured over
each impression, creating ceramic part patterns. Once the ceramic had
cured, the rubber was withdrawn. Aluminum inserts were then cast
from the ceramic part patterns, finished and mounted to produce a final
assembly tool. Within days of the tool's completion date, Santin quickly
shot over 500 parts in a variety of colors, and most importantly, in the
real production material, ABS plastic.

These production quality parts (see Figure 3.5) were delivered to
Bose exactly four weeks after the day they placed the order. The plastic
parts were mounted into several of the cars and successfully color
matched as well as road tested for design, function, and durability. By
using the Shaw process, Santin shaved five weeks off the tooling cycle,
a time savings of over 50%. "Even with an open schedule and every
single man-hour dedicated to this project, we still would not have been
able to meet our deadline using traditional tooling," says Santin.

Figure 3.5: Assembly tool and car speaker grills made using SLA (Courtesy 3D Systems)

3.1.7   Research and Development

3D Systems' research focus is on improving process and developing new materials, and application, especially rapid tooling.

3D Systems and Ciba-Geigy Ltd. are in a joint research and development program, continually working on new resins which have better mechanical properties, are faster and easier to process, and are able to withstand higher temperatures [12].

One of the other most important areas of research is in rapid tooling, i.e., the realization of prototype molds and ultimately production tooling inserts [13]. 3D Systems is involved in 15 cooperative rapid tooling partnerships with various industrial, university and government agencies. Methods studied, tested or being evaluated include those used for soft and hard tooling. With the purchase of 3D Keltool in September 1996, 3D Systems now has a means for users to go from a CAD model, to a SL master to Keltool cores and cavities capable of producing in excess of one million injection molded parts in a wide range of engineering thermoplastics such as polypropylene, nylon, ABS, polyethylene and polycarbonate.

3.1.8   Others

3D Systems is by far the largest supplier of RP systems worldwide.

3.2   CUBITAL'S SOLID GROUND CURING (SGC)

3.2.1   Company

The Solid Ground Curing (SGC) System is produced by Cubital Ltd. and its address is Cubital Ltd., 13 Hasadna St., P.O.B. 2375, Industrial Zone North Raanana, 43650 Israel. Outside Israel, Cubital America Inc. is located at 1307F Allen Drive Troy, MI 48083, USA and Cubital GmbH, at Ringstrasse 132 55543 Bad-Kreuznach, Germany. Cubital Ltd.'s operations began in 1987 as a spin-off from Scitex Corporation and commercial sales began in 1991.

3.2.2 Products

3.2.2.1 *Models and Specifications*

Cubital's products include the Solider 4600 and Solider 5600. The Solider 4600 is Cubital's entry level three-dimensional model making system based on Solid Ground Curing. The Solider 5600, Cubital's sophisticated high-end system, provides a wider range and options for the varied modeling demands of Solid Ground Curing. Table 3.2 summarizes the specifications of the two machines.

Table 3.2: Cubital Inc.'s Solider 4600 and 5600 (Source from Cubital Inc.)

Model	Solider 4600	Solider 5600
Irradiation medium	High power UV lamp	
XY resolution (mm)	Better than 0.1	
Surface definition (mm)	0.15	0.15
Elevator vertical resolution (mm)	0.15	0.1–0.2
Minimum feature size (mm)	0.4 (horizontal, X–Y) 0.15 (vertical, Z)	0.4 (horizontal, X–Y) 0.15 (vertical, Z)
Work volume, XYZ (mm × mm × mm)	$350 \times 350 \times 350$	$500 \times 350 \times 500$
Production rate (cm³/hr)	550	1311
Minimum layer thickness (mm)	0.06	0.06
Dimensional accuracy	0.1%	0.1%
Size of unit, XYZ (m × m × m)	$1.8 \times 4.2 \times 2.9$	$1.8 \times 4.2 \times 2.9$
Data control unit	Data Front End (DFE) workstation	
Power supply	380–415 V_{AC}, 3 phase, 50 kW	380–415 V_{AC}, 3 phase, 50 kW

Cubital's system uses several kinds of resins, including liquid resin and cured resin as materials to create parts, water soluble wax as support material and ionographic solid toner for creating an erasable image of the cross-section on a glass mask.

3.2.2.2 *Advantages and Disadvantages*

The Solider system has the following advantages:

(1) *Parallel processing.* The process is based on instant, simultaneous curing of a whole cross-sectional layer area (rather than point-by-point curing). It has a high speed throughput that is about eight times faster than its competitors. Its production costs can be 25% to 50% lower. It is a time and cost saving process.

(2) *Self-supporting.* It is user-friendly, fast, and simple to use. It has a solid modeling environment with unlimited geometry. The solid wax supports the part in all dimensions and therefore a support structure is not required.

(3) *Fault tolerance.* It has good fault tolerances. Removable trays allow job changing during a run and layers are erasable.

(4) *Unique part properties.* The part that the Solider system produces is reliable, accurate, sturdy, machinable, and can be mechanically finished.

(5) *CAD to RP software.* Cubital's RP software, Data Front End (DFE), processes solid model CAD files before they are transferred to the Cubital's machines. The DFE is an interactive and user-friendly software.

(6) *Minimum shrinkage effect.* This is due to the full curing of every layer.

(7) *High structural strength and stability.* This is due to the curing process that minimizes the development of internal stresses in the structure. As a result, they are much less brittle.

(8) *No hazardous odors are generated.* The resin stays in a liquid state for a very short time, and the uncured liquid is wiped off immediately. Thus safety is considerably higher.

The Solider system has the following disadvantages:

(1) *Requires large physical space.* The size of the system is much larger than other systems with a similar build volume size.
(2) *Wax gets stuck in corners and crevices.* It is difficult to remove wax from parts with intricate geometry. Thus, some wax may be left behind.
(3) *Waste material produced.* The milling process creates shavings, which have to be cleaned from the machine.
(4) *Noisy.* The Solider system generates a high level of noise as compared to other systems.

3.2.3   Process

The Cubital's Solid Ground Curing process includes three main steps: data preparation, mask generation and model making [14].

3.2.3.1   *Data Preparation*

In this first step, the CAD model of the job to be prototyped is prepared and the cross-sections are generated digitally and transferred to the mask generator. The software used, Cubital's Solider DFE (Data Front End) software, is a motif-based special-purpose CAD application package that processes solid model CAD files prior to sending them to Cubital Solider system. DFE can search and correct flaws in the CAD files and render files on-screen for visualization purposes. Solider DFE accepts CAD files in the STL format and other widely used formats exported by most commercial CAD systems.

3.2.3.2   *Mask Generation*

After data are received, the mask plate is charged through an "image-wise" ionographic process (see item 1, Figure 3.6). The charged image is then developed with electrostatic toner.

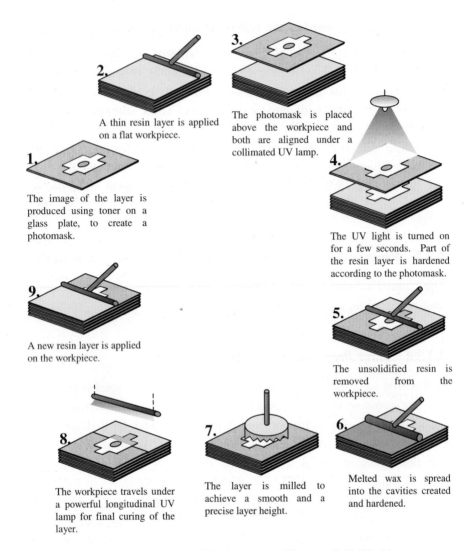

1. The image of the layer is produced using toner on a glass plate, to create a photomask.

2. A thin resin layer is applied on a flat workpiece.

3. The photomask is placed above the workpiece and both are aligned under a collimated UV lamp.

4. The UV light is turned on for a few seconds. Part of the resin layer is hardened according to the photomask.

5. The unsolidified resin is removed from the workpiece.

6. Melted wax is spread into the cavities created and hardened.

7. The layer is milled to achieve a smooth and a precise layer height.

8. The workpiece travels under a powerful longitudinal UV lamp for final curing of the layer.

9. A new resin layer is applied on the workpiece.

Figure 3.6: Solid Ground Curing process (Courtesy Cubital Ltd.)

3.2.3.3   *Model Making*

In this step, a thin layer of photopolymer resin is spread on the work surface (see item 2, Figure 3.6). The photo mask from the mask generator is placed in close proximity above the workpiece, and aligned under a collimated UV lamp (item 3). The UV light is turned on for a few seconds (item 4). The part of the resin layer which is exposed to the UV light through the photo mask is hardened. Note that the layers laid down for exposure to the lamp are actually thicker than the desired thickness. This is to allow for the final milling process. The un-solidified resin is then collected from the workpiece (item 5). This is done by vacuum suction. Following that, melted wax is spread into the cavities created after collecting the liquid resin (item 6). Consequently, the wax in the cavities is cooled to produce a wholly solid layer. Finally, the layer is milled to its exact thickness, producing a flat solid surface ready to receive the next layer (item 7).

In the SGC 5600, an additional step (item 8) is provided for final curing of the layer whereby the workpiece travels under a powerful longitudinal UV lamp. The cycle repeats itself until the final layer is completed.

The main components of the Solider system are (see Figure 3.7):

(1) Data Front End (DFE) workstation.
(2) Model Production Machine (MPM). It includes:
   (i) Process engine,
   (ii) Operator's console,
   (iii) Vacuum generator.
(3) Automatic Dewaxing Machine (optional).

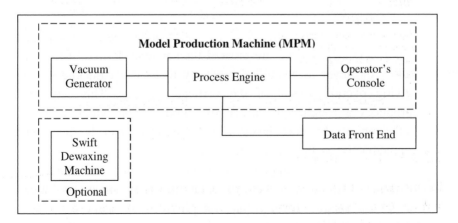

Figure 3.7: Solider system block diagram (Courtesy Cubital Ltd.)

3.2.4   Principle

Cubital's RP technology creates highly physical models directly from computerized three-dimensional data files. Parts of any geometric complexity can be produced without tools, dies or molds by Cubital's RP technology.

The process is based on the following principles:

(1) Parts are built, layer by layer, from a liquid photopolymer resin that solidifies when exposed to UV light. The photopolymerization process is similar to that described in Section 3.1.4, except that the irradiation source is a high power collimated UV lamp and the image of the layer is generated by masked illumination instead of optical scanning of a laser beam. The mask is created from the CAD data input and "printed" on a transparent substrate (the mask plate) by an nonimpact ionographic printing process, a process similar to the Xerography process used in photocopiers and laser printers [15]. The image is formed by depositing black powder, a toner which adheres to the substrate electrostatically. This is used to mask the uniform illumination of the UV lamp. After exposure, the electrostatic toner is removed from the substrate for reuse and the pattern for the next layer is similarly "printed" on the substrate.

(2) Multiple parts may be processed and built in parallel by grouping them into batches (runs) using Cubital's proprietary software.

(3) Each layer of a multiple layer run contains cross-sectional slices of one or many parts. Therefore, all slices in one layer are created simultaneously. Layers are created thicker than desired. This is to allow the layer to be milled precisely to its exact thickness, thus giving overall control of the vertical accuracy. This step also produces a roughened surface of cured photopolymer, assisting adhesion of the next layer to it. The next layer is then built immediately on the top of the created layer.

(4) The process is self-supporting and does not require the addition of external support structures to emerging parts since continuous structural support for the parts is provided by the use of wax, acting as a solid support material.

3.2.5 Applications

The applications of Cubital's system can be divided into four areas:

(1) *General applications.* Conceptual design presentation, design proofing, engineering testing, integration and fitting, functional analysis, exhibitions and pre-production sales, market research, and inter-professional communication.
(2) *Tooling and casting applications.* Investment casting, sand casting, and rapid, tool-free manufacturing of plastic parts.
(3) *Mold and tooling.* Silicon rubber tooling, epoxy tooling, spray metal tooling, acrylic tooling, and plaster mold casting.
(4) *Medical imaging.* Diagnostic, surgical, operation and reconstruction planning and custom prosthesis design.

3.2.6 Examples

3.2.6.1 *Cubital Prototyping Machine Builds Jeep in 24 Hours*

Toledo Model and Die Inc. (TMD) of Toledo, Ohio, has used Cubital's Solider 5600 rapid prototyping system to produce three design iterations of a toy jeep (Figure 3.8) in three days [16]. The plastic push toy is about 30 cm (12 in) long, 23 cm (9 in) wide and 23 cm (9 in) high.

Figure 3.8: The all-white Cubital prototype of the toy jeep beside the final test prototype cast from it in urethane reproductions (Courtesy Cubital Americal Inc.)

Using the Cubital Solider 5600, TMD was able to produce each design iteration in 13 hours of machine time. By using the Cubital front-end software, the parts were "nested" within a working volume only 10 cm (4 in) deep, within the Solider's working area 50 cm × 35 cm (20 in by 14 in). Initial file checking and post-production wax removal accounted for the rest of the 24 hours.

3.2.6.2   *New Cubital-based Solicast Process Offers Inexpensive Metal Prototypes in Two Weeks*

Schneider Prototyping GmbH created a metal prototype (see Figure 3.9) for investment cast directly from CAD files in two weeks with SoliCast [17]. Starting with a CAD file, Schneider can deliver a metal prototype in two weeks, whereas similar processes with other rapid prototyping methods take four to six weeks and conventional methods based on CNC prototyping can take 10 to 16 weeks. In general, Schneider's prototypes can be as much as 50% cheaper than those produced by other rapid prototyping methods.

Figure 3.9: Cubital prototypes and metal prototypes created from them
(Courtesy Cubital America Inc.)

3.2.6.3   *Cubital Saves the Day — In Fact, About Six Weeks — For Pump Developers*

Engineers and technicians in the production engineering department of Sintef, a research and development firm in Trondheim, Norway, made an accurate, full scale plastic model of the impeller (Figure 3.10), which measures about 220 mm in diameter and 50 mm high, using their Solider 5600 [18]. This model was used to create an investment casting pattern that was used to cast a prototype impeller in metal. From the time the engineers in Sintef got the CAD files till they had the investment casting pattern was three weeks — about a third the time it would have taken using CNC tools.

Figure 3.10: Plastic model of impeller for new pump (Courtesy Cubital America Inc.)

3.2.7   Research and Development

Cubital is doing research on faster processing, higher performance, higher resolution graphics, smoother and more accurate renderings and shadings.

Following the increasing demand for RP parts with improved mechanical characteristics, also for direct utilization of RP parts in investment casing and rapid tooling processes, Cubital is developing an extension to the Solid Ground Curing process that will be upgradable on standard Solider Systems. The Extended Process will enable production of parts made of enhanced thermoset, thermoplastic and metallic materials, significantly increasing the end users' ability to select mechanical properties suitable for specific applications or production processes. Particularly, parts made of castable wax can be used directly for investment casting, and parts produced of Metal Sprayed Zinc can be used directly for plastic injection molding.

3.2.8   Others

Cubital's Solider systems have been sold worldwide in North America, Asia and Europe.

3.3   D-MEC'S SOLID CREATION SYSTEM (SCS)

3.3.1   Company

The Solid Creation System (SCS) has been jointly developed by Sony Corporation, JSR Corporation and D-MEC Corporation. The software and hardware have been created by Sony Corporation, the UV curable resin by JSR (Japan Synthetic Rubber) Corporation and the forming and applied technology by D-MEC Corporation. D-MEC was established on 28 Feb. 1990. The address of the company is JSR Building, 2-11-24 Tsukiji, Chuo-ku, Tokyo 104-0045, Japan.

3.3.2   Products

Based on the principle of laser cured polymer using layer manufacturing, D-MEC is the first company to offer a 1/2 meter cubic tank size and with scanning speed of up to 5 m/s.

In cooperation with JSR which developed the resin used in the system, the Solid Creator was well received in Japan, especially amongst auto makers and electronics industries.

Figure 3.11: Solid Creation System SCS-1000 HD (Courtesy D-MEC Corporation)

Table 3.3: Solid Creator model specifications (Courtesy D-MEC Corporation)

Model	SCS-1000HD	SCS-2000	SCS-3000	SCS-8000
Laser	He–Cd	Solid-state semiconductor laser		
Modulating device	AOM (Opto-acoustic element)			
Deflector device	Galvanometer mirror system (defocus via sweep and straightening function)			
Spot size (mm)	0.05–0.3 (auto adjustment)	0.15–0.8 (auto adjustment)		
XY sweep speed (m/sec)	2 (max)	10 (max)		
Work volume, XYZ (mm × mm × mm)	300 × 300 × 300	600 × 500 × 500	1000 × 800 × 500	600 × 500 × 500
Vat capacity (liter)	45 (ex-changeable)	265 (ex-changeable)	840 (ex-changeable)	265 (ex-changeable)
Size of unit, XYZ (mm × mm × mm)	1425 × 1100 × 1590	1650 × 1400 × 1820	2340 × 1700 × 2840	1650 × 1400 × 1820
Date control unit	Windows NT			
Power supply unit	AC 100 V 10 A	AC 200 V 30 A		

3.3.2.1 *Models and Specifications*

D-MEC's products include four models: Solid Creator SCS-1000HD (see Figure 3.11), SCS-2000, SCS-3000 and SCS-8000. The specifications of these machines are summarized in the Table 3.3.

3.3.2.2 *Advantages and Disadvantages*

Solid Creator has the following strengths:

(1) *Large build volume.* The tank size is large and large prototypes (especially large full-scale prototypes) can be produced.
(2) *Accurate.* High accuracy (0.04 mm repeatability) model may be produced.

Solid Creator has the following weaknesses:

(1) *Requires support structures.* Structures that have overhangs and undercuts must have supports that are designed and fabricated together with the main structure.
(2) *Requires post-processing.* Post-processing includes removal of supports and other unwanted materials, which is tedious, time-consuming and can damage the model.
(3) *Requires post-curing.* Post-curing may be needed to cure the object completely and ensure the integrity of the structure.

3.3.3 Process

Solid Creator creates the three-dimensional model by laser curing polymer layer by layer. Its process comprises five steps: generating the CAD model, slicing the CAD model and transferring data, scanning the resin surface, lowering the elevator, completion of the prototype and post-processing.

First, a CAD model, usually a solid model, is created in a commercial CAD system, like CADDS5, CATIA or Pro/Engineer. Three-dimensional CAD data of the part from the CAD system are converted to the sliced cross-sectional data which the SCS will use in creating the solid. This is the slicing process. Editing may be necessary if the slicing is not carried out properly. Both the slicing and editing processes can be done either ON- or OFF-line. Consequently, the section-data are passed to the laser controller for the UV curing process.

The ultraviolet laser then scans the resin surface in the tank to draw the cross-sectional shape based on these data. The area of the resin surface which is hit by the laser beam is cured, changing from liquid to solid on the elevator.

The elevator descends to allow the next solid layer to be created by the same process. This is repeated continuously to laminate the necessary number of thin cross-sectional layers to shape the three-dimensional part. Finally, when the model is completed, the elevator is raised and the model is lifted out and post-curing treatment is applied.

The main hardware of the Solid Creator includes:

(1) The Sony NEWS UNIX workstation.
(2) The main machine controller (VME based, MTOS — Multitasking Operating System).
(3) The galvano mirror and its controller.
(4) The optical system including the laser, the lens and the Acoustic Optical Modulator (AOM).
(5) The photopolymer tank.
(6) The elevator mechanism.

The main software of the Solid Creator consists of two parts: the slice data generator which uses inputs from various CAD systems and creates the cross-sectional slices and the editing software for slice data which includes the automatic support generation software. The software used to control the Solid Creator also uses principles of man–machine interfaces.

3.3.4 Principle

D-MEC's Solid Creator rapid prototyping system is based on the principle of polymer curing by exposing to ultraviolet light and manufacturing by layering. The basic principles and techniques used are similar to that described in Section 3.1.4. However, D-MEC called this the "Optical Shaping Method".

In the process, parameters which affect performance and functionality of the machine are scanning pitch, step period, step size, scanner delay, jump size, jump delay, scanning pattern and the resin's properties.

3.3.5   Applications

The general application areas are given as follows:

(1) Mock-up in product design.
(2) Design study and sales sample of new products.
(3) Use as parts without need of modification in small lot production.
(4) Simplified mold tool and master for investment casting and other similar kinds of processes.

D-MEC Corporation has used Solid Creator mainly to design and prototype its own products like the drum base for video tape recorder. By using Solid Creator, D-MEC is able to ascertain optimum design in its products and important functional tests are carried out on the prototype as well.

When Solid Creator is used for medical purpose, surgical operation time can be significantly shortened by checking the defective or diseased areas using models reconstructed on the system from other frequently used computer images, for example, CT Scan, etc.

3.3.6   Others

Solid Creator machines have been installed throughout the world, though mainly in Japan. Besides these, D-MEC provides comprehensive bureau services to industries in Japan.

3.4   CMET'S SOLID OBJECT ULTRAVIOLET-LASER PRINTER (SOUP)

3.4.1   Company

CMET (Computer Modeling and Engineering Technology) Inc. was established in November 1990 with the Mitsubishi Corporation as the major share holder and two other substantial shareholders, NTT Data Communication Systems and Asahi Denak Kogyo K.K. However, in 2001, CMET Inc. merged with Opto-Image Company, Teijin Seiki Co. Ltd. The company's address is CMET Inc., Kamata Tsukimura Building, 5-15-8, Kamata, Ohta-ku, Tokyo 144-0052, Japan.

3.4.2   Products

3.4.2.1   *Models and Specifications*

CMET claimed to have 50% of Japan's market share in RP machines (March, 1994) [19]. The systems use a galvanometer mirror with a Z focus unit. There are three models of machines: SOUP II 600GS-02, SOUP II 600GS-05 and SOUP II 600GS-10. Figure 3.12 shows a typical SOUP II 600GS machine. The material used by the SOUP system is a photo-curable epoxy resin.

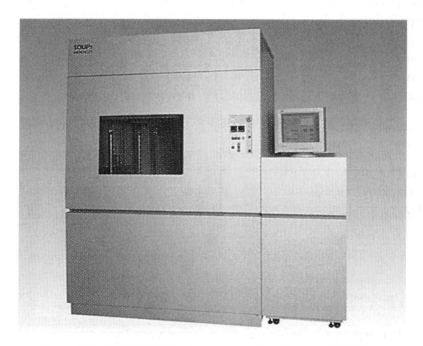

Figure 3.12: SOUPII 600GS printer type machine (Courtesy CMET Inc.)

The galvanometer mirror type uses an ultra-high speed scan for raster scanning with a galvanometer mirror, creating a part in a very short time, with scanning speeds of up to 20 m/s for the SOUPII 600GS series. These types of SOUP machines can make the laser beam diameter very small (smallest, 0.1 mm) for creating an extremely small model with high precision. The central working section provides a high position precision suited to very small models like connectors, and solid structure models like components. In addition, the use of a galvano-meter mirror makes the unit design compact. A summary table of the machines available is given in Table 3.4.

Table 3.4: SOUP's models and specifications (Courtesy CMET Inc.)

Model	SOUP II 600GS-02	SOUP II 600GS-05	SOUP II 600GS-10
Laser type	Solid State		
Laser power (mW)	200	500	1000
Scanning system	Galvanometer mirror with Z focus unit		
Max. scanning speed (m/s)	20		
Beam diameter	Variable 0.1–0.5 mm		
Max. build envelope, XYZ (mm × mm × mm)	$600 \times 600 \times 500$		
Minimum build layer (mm)	0.05		
Vat	Interchangeable		
Recoating system	(1) New recoating system (2) Roll coater system (option)		
Resin level controller	Level adjustment block axis with high accuracy level sensor		
Power supply	100 V_{AC}, single phase, 20 A (Control unit) & 30 A (Heater)		
Size of unit, XYZ (mm × mm × mm) – Process module – Control module	$1575 \times 970 \times 2160$ $610 \times 1020 \times 1230$		
Weight (kg)	1700		
Data control unit	Win 2000		

3.4.2.2 *Advantages and Disadvantages*

The main advantages of SOUP systems are:

(1) *New recoating system.* The new recoating system provides a more accurate Z layer and shorter production time.
(2) *Software system.* The software system allows for real time processing.
(3) *High scanning speed.* It has scanning speeds of up to 20 m/sec.

The main disadvantages of SOUP systems are:

(1) *Requires support structures.* Structures that have overhangs and undercuts must have supports that are designed and fabricated together with the main structure.
(2) *Requires post-processing.* Post-processing includes removal of supports and other unwanted materials, which is tedious, time-consuming and can damage the model.
(3) *Requires post-curing.* Post-curing may be needed to cure the object completely and ensure the integrity of the structure.

3.4.3 Process

The SOUP process contains the following three main steps:

(1) *Creating a 3D model with a CAD system*: The three-dimensional model, usually a solid model, of the part is created with a commercial CAD system. Three-dimensional data of the part are then generated.
(2) *Processing the data with the SOUPware*: Often data from a CAD system are not faultless enough to be used by the RP system directly. SOUP's software, SOUPware, can edit CAD data, repair its defects, like gaps, overlaps, etc. (see Chapter 6), slice the model into cross-sections and finally, SOUPware generates the corresponding SOUP machine data.
(3) *Making the model with the SOUP units*: The laser scans the resin, solidifying it according to the cross-sectional data from SOUPware. The elevator lowers and the liquid covers the top layer of the part

which is recoated and prepared for the next layer. This is repeated until the whole part is created.

SOUP's hardware contains a communication controller, a laser controller, a shutter/filter controller, a scanner controller (galvanometer mirror unit) and an elevator controller. The SOUPware, SOUP's software is a real-time working, multi-user and multi-machine control software. It has functions such as simulation, convenient editor for editing and error repair, 3D offset, loop scan for filling area between the outlines, delectable structure and automatic support generation.

3.4.4   Principle

The SOUP system is based on the laser lithography technology (Figure 3.13) which is similar to that described in Section 3.1.4. The one major difference is in the optical scanning system. The SOUP system has the option for *XY* plotter, which is easier to control and has less optic problems than the galvanometer mirror system. The main trade-off is in scanning speed and consequently, the building speed. Parameters which influence performance and functionality are

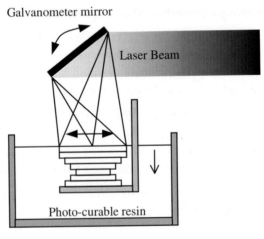

Figure 3.13: Schematic of SOUP® system (Adapted from CMET brochure)

galvanometer mirror precision for the galvanometer mirror machine, laser spot diameter, slicing thickness and resin properties.

3.4.5 Applications

The application areas of the SOUP systems include the following:

(1) Concept models for design verification, visualization and commercial presentation purposes.
(2) Working models for form fitting and simple functional tests.
(3) Master models and patterns for silicon molding, lostwax, investment casting, and sand casting.
(4) As completely finished parts.
(5) Medical purposes. Creating close to exact physical models of a patient's anatomy from CT and MRI scans.
(6) Three-dimensional stereolithography copy of existing product.

3.4.6 Research and Development

CMET focuses its research and development on a new recoating system. The new recoating system can make the Z layer with high accuracy even though a small Z slice pitch (0.05 mm) is used. As the down-up process of the Z table is not productive, the "no-scan" time can be shortened by more than 50%. More functions will be added into CMET's software — SOUPware. CMET is also conducting the following experiments for improving accuracy:

(1) Investigation on relationship between process parameters and curl distortion: this experiment is done by changing laser power, scan speed, and fixing other parameters like Z layer pitch and beam diameter.
(2) Investigation on effects of scan pattern: this experiment aims to compare building results between using and not using scanning patterns.
(3) Comparative testing of resins' properties on distortion.

3.5   TEIJIN SEIKI'S SOLIFORM SYSTEM

3.5.1   Company

Teijin Seiki Co. Ltd., originally founded in 1944, is a diversified industrial technology company supporting and supplying a wide range of industrial components. Its Soliform system is based on the SOMOS (Solid Modeling System) laser curing process developed by Du Pont Imaging Systems. The company acquired exclusive Asian rights to the equipment aspects of this technology in 1991 under its Solid Imaging Department, with its first machine shipped in 1992. Since 2001, Teijin Seiki has merged with CMET Inc. under one umbrella. The CMET's company address is Kamata Tsukimura Building, 5-15-8, Kamata, Ohta-ku, Tokyo 144-0052, Japan.

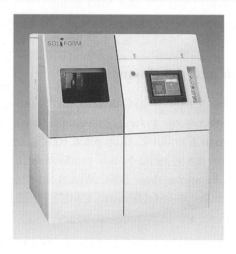

Figure 3.14: The Soliform 250B solid forming system (Courtesy Teijin Seiki Co. Ltd.)

3.5.2   Product

3.5.2.1   *Models and Specifications*

Teijin Seiki Co. Ltd. produces two main series (250 and 500 series) of the Soliform system. The Soliform 250 series machine is shown in Figure 3.14.

Table 3.5: System specification of Soliform (Courtesy Teijin Seiki Co. Ltd.)

Model	SOLIFORM 250B	SOLIFORM 250EP	SOLIFORM 500C	SOLIFORM 500EP
Laser type	Solid state		Solid state	
Laser power (mW)	200	1000	400	1000
Scanning system	Digital scanner mirror		Digital scanner mirror	
Maximum scanning speed (m/s)	12 (max)		24 (max)	
Beam diameter	Fixed: 0.2 mm Variable: 0.1–0.8 mm (option)		Fixed: 0.2 mm Variable: 0.2–0.8 mm (option)	
Max. build envelope, XYZ (mm × mm × mm)	$250 \times 250 \times 250$		$500 \times 500 \times 500$	
Minimum build layer (mm)	0.05		0.05	
Vat	Interchangeable		Interchangeable	
Recoating system	Dip coater system		Dip coater system	
Resin level controller	Counter volume system		Counter volume system	
Power supply	100 V_{AC}, single phase, 30 A		200 V_{AC}, 3 phase, 30 A	
Size of unit, XYZ (mm × mm × mm)	$1430 \times 1045 \times 1575$		$1850 \times 1100 \times 2280$	
Weight (kg)	400		1200	

The specifications of the machines are summarized in Table 3.5. The materials used by Soliform are primarily SOMOS photopolymers supplied by Du Pont and TSR resins, its own developed resin.

3.5.2.2   *Advantages and Disadvantages*

The Soliform system has several technological advantages:

(1) *Fast and accurate scanning.* Its maximum scanning speed of 24 m/s is faster than all other RP systems. Its accurate scan system is controlled by digital encoder servomotor techniques.

(2) *Good accuracy.* It has exposure control technology for producing highly accurate parts.
(3) *Photo resins.* It has a wide range of acrylic-urethane resins for various applications.

The Soliform system has the following disadvantages:

(1) *Requires support structures.* Structures that have overhangs and undercuts must have supports that are designed and fabricated together with the main structure.
(2) *Requires post-processing.* Post-processing includes removal of supports and other unwanted materials, which is tedious, time-consuming and can damage the model.
(3) *Requires post-curing.* Post-curing may be needed to cure the object completely and ensure the integrity of the structure.

3.5.3   Process

The process of the Soliform comprises the following steps: concept design, CAD design, data conversion, solid forming and plastic model.

(1) *Concept design*: Product design engineers create the concept of the product, i.e., the concept design. This may or may not necessarily be done on a computer.
(2) *CAD design*: The three-dimensional CAD model of the concept design is created in the SUN-workstation.
(3) *Data conversion*: The three-dimensional CAD data is transferred to the Soliform software to be tessellated and converted into the STL file.
(4) *Solid forming*: This is the part building step. A solid model is formed by the ultraviolet laser layer by layer in the machine.
(5) *Plastic model*: This is the post-processing step in which the completed plastic part is cured in an oven to be further hardened.

The Soliform system contains the following hardware: a SUN-EWS workstation, an argon ion laser, a controller, a scanner to control the laser trace of scanning and a tank which contains the photopolymer resin.

3.5.4   Principle

The Soliform creates models from photo-curable resins based essentially on the principles described in Section 3.1.4. The resin developed by Teijin is an acrylic-urethane resin with a viscosity of 40 000 centapoise and a flexural modulus of 52.3 MPa as compared to 9.6 MPa for a grade used to produce conventional prototype models.

Parameters which influence performance and functionality are generally similar to those described in Section 3.1.4, but for Soliform, the properties of the resin and the accuracy of the laser beam are considered more significant.

3.5.5   Applications

The Soliform has been used in many areas, such as injection molding (low cost die), vacuum molding, casting and lost wax molding.

(1) *Injection molding (low cost die).* The Soliform can be used to make an injection molding or low cost dies based on a process described in Figure 3.15.

Comparing this method with the conventional processes, this process has shorter development and execution time. The created CAD data can be used for a mass-production die and no machining is necessary. Photopolymers that can be used include SOMOS-2100 and SOMOS-5100. An example of the product molded in ABS is shown together with the injection mold and the SOLIFORM pattern in Figure 3.16.

(2) *Vacuum molding.* Vacuum molding molds are made by the Soliform using a process similar to that illustrated in Figure 3.15. The photopolymers that can be used include SOMOS-2100, SOMOS-3100 and SOMOS-5100.

(3) *Casting.* The process of making a casting part form is again similar to the process illustrated in Figure 3.15 with the exception of using casting sand instead of the injection mold. The photopolymer recommended is SOMOS-3100.

(4) *Lost wax molding.* Again this process is similar to the one described in Figure 3.15. The main difference is that the wax

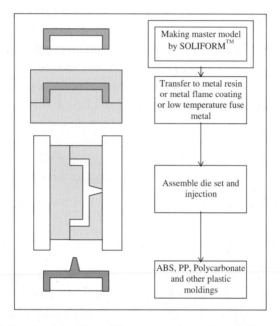

Figure 3.15: Creating injection molding parts (Adapted from Teijin Seiki Co. Ltd.)

Figure 3.16: RP model (left), the ABS injection molding (center) and the
low cost metal die (right) (Courtesy Teijin Seiki Co. Ltd.)

pattern is made and later burnt out before the metal part is molded
afterwards. The recommended photopolymer is SOMOS-4100.

(5) *Making injection and vacuum molding tools directly.* The other
    application of SOLIFORM is the process of making the tools for
    injection molding and vacuum molding directly. From the CAD

data of the part, the CAD model of the mold inserts are created. These new CAD data are then transferred the usual way to SOLIFORM to create the inserts. The inserts are then cleaned and post-cured before the gates and ejector pin holes are machined. They are then mounted on the mold sets and subsequently on the injection molding machine to mold the parts.

Figure 3.17 describes schematically the process flow. Figures 3.18 and 3.19 show the injection mold cavity insert and the molding for a

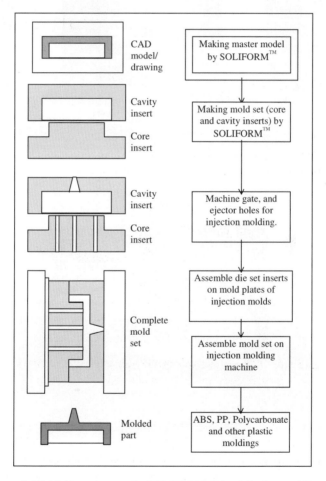

Figure 3.17: Making process of a SOLIFORM direct injection molding die

Figure 3.18: Core and cavity mold inserts on molding machine of the handy telephone
(Courtesy Teijin Seiki Co. Ltd.)

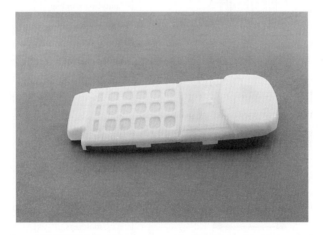

Figure 3.19: Injection molded part from the SOLIFORM molds of the handy telephone
(Courtesy Teijin Seiki Co. Ltd.)

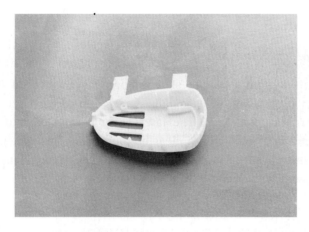

Figure 3.20: Direct vacuum molded mouse cover (Courtesy Teijin Seiki Co. Ltd.)

handy telephone respectively. In this example, the injection material is ABS resin, the mold clamping force is 60 tons, the molding quantity is 22 pieces and the molding cycle time is one minute per shot.

A process similar to that described in Figure 3.17 can also be used for creating molds and dies for vacuum molding. Figure 3.20 shows a mouse cover that is molded using vacuum molding from tools created by SOLIFORM.

3.5.6   Research and Development

The research on expanding Soliform's applications with better resins and improved laser accuracy and precision are being carried out at Teijin Seiki.

3.6   AUTOSTRADE'S E-DARTS

3.6.1   Company

Autostrade Co. Ltd., founded in 1984, is a Japanese manufacturer of stereolithography systems. Its address is at 13-54 Ueno-machi, Oita-City, Oita 870-0832 Japan. Its first machine was sold in 1998.

3.6.2   Products

3.6.2.1   *Models and Specifications*

The E-Darts system uses the stereolithography method of hardening liquid resin by a beam of laser light. The set-up of the laser system is found under the resin tank. The laser beam is directed into the resin which is in a container with a clear bottom plate. This system uses an acrylic resin to produce models. Table 3.6 lists the specifications of the E-darts system while Figure 3.21 shows the E-Darts system.

Table 3.6: Specifications of E-Darts machine

Model	E-Darts
Laser type	Semiconductor
Laser power (mW)	30
Spot size (mm)	0.1
XY sweep speed (m/s)	0.03
Elevator vertical resolution (mm)	0.05
Work volume, XYZ (mm × mm × mm)	$200 \times 200 \times 200$
Maximum part weight (kg)	No count
Minimum layer thickness (mm)	0.05
Size of unit, XYZ (mm × mm × mm)	$430 \times 500 \times 515$
Data control unit	Win 98/ME
Power supply	12 V

Figure 3.21: E-Darts system

3.6.2.2 Advantages and Disadvantages

The E-Darts system has the following strengths:

(1) *Low price.* The price of the E-Darts system is 2 980 000 yen (US$27 000), which is lower than most other systems.
(2) *Low operating cost.* The introduction of an improved fluid surface regulation system has decreased the necessary volume of resin, which actually eliminates the need to stock the resin. This in turn reduces the running cost of the system.
(3) *Compact size.* The system size is 430 mm × 500 mm × 515 mm.
(4) *Portable.* The weight of the system is less than 25 kg.
(5) *Ease of installation.* The set-up time for the system is under 1 hr.

The E-Darts system has the following weaknesses:

(1) *Requires support structures.* Structures that have overhangs and undercuts must have supports that are designed and fabricated together with the main structure.
(2) *Requires post-processing.* Post-processing includes removal of supports and other unwanted materials, which is tedious, time-consuming and can damage the model.
(3) *Requires post-curing.* Post-curing may be needed to cure the object completely and ensure the integrity of the structure.

3.6.3 Process

The software comprises of two components. First, the STL file editing software changes the CAD data into slice data. Second, the controller software drives the hardware. Both of these components use the Windows Operating System.

3.6.4 Principle

The model is formed by liquid resin which is cured by laser light beamed from below the resin chamber. While 3D Systems' SLA uses a platform which is dipped into the resin tank when one layer is completed, E-Darts system uses another method. The platform or

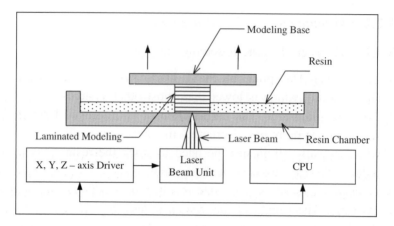

Figure 3.22: E-Darts process

modeling base (Figure 3.22) is raised upwards each time a layer is completed. Otherwise, the E-Darts is based on the laser lithography technology which is similar to that described in Section 3.1.4.

3.6.5 Research and Development

Autostrade is working on improving the compactness and portability of their product.

3.7 MEIKO'S RAPID PROTOTYPING SYSTEM FOR THE JEWELRY INDUSTRY

3.7.1 Company

Meiko Co. Ltd., a company making machines for analyzing components of gas such as CO_2, NO_X and O_2, factory automation units such as loaders and unloaders for assembly lines, and two arm robots, was founded in June 1962. The system for jewelry industry was developed together with Yamanashiken-Industrial Technical Center in the early 1990s. The company's address is Meiko Co. Ltd., Futaba Industrial Park, Shimoimai 732, Futaba-cho, KitaKoma-gun, 407-0105, Yamanashi-ken, Japan.

3.7.2 Product

3.7.2.1 *Model and Specifications*

Meiko's LC-510 (see Figure 3.23) is an optical modeling system that specializes in building prototypes for small models such as jewelry. The system is named MEIKO. The specifications of the LC-510 are summarized in Table 3.7.

The system uses UVM-8001, a photo-curable resin jointly developed by Meiko Co. Ltd. as the material to create small models. The consumable components in the system are the He–Cd laser tube and the photo-curable resin. The computer and the CAD system, which is named JCAD3/Takumi, shown in Figure 3.23 are options.

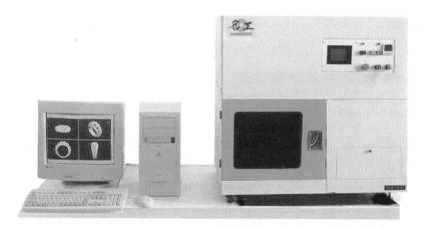

Figure 3.23: LC-510 optical modeling and CAD system (Courtesy Meiko Co. Ltd.)

3.7.2.2 *Advantages and Disadvantages*

The strengths of the product and process are as follows:

(1) *Good accuracy.* Highly accurate modeling that is extremely suitable for jewelry.
(2) *Low cost.* Relatively inexpensive cost for prototyping.
(3) *Cost saving.* Lead time is cut and cost is saved by labor saving.

Table 3.7: Specifications of the LC-510 (Courtesy Meiko Co. Ltd.)

Model	LC-510
Laser type	He–Cd
Laser power (mW)	5
Spot size (mm)	0.08
Method of laser operation	X–Y plotter
Elevator vertical resolution (mm)	0.01
Work volume, XYZ (mm × mm × mm)	$100 \times 100 \times 60$
XY repeatability	±0.01
Minimum layer thickness (mm)	0.01
Size of unit, XYZ (mm × mm × mm)	$1.0 \times 0.7 \times 1.38$
Data control unit	Windows 98/2000/NT
Power supply	100 V_{AC} 15 A
Price (USD, 2001)	Open price (the flexibility is given to the various agents selling the machines)

(4) *Exclusive CAM software.* The CAM software is for jewelry and small parts.

(5) *Ease of manufacturing complicated parts.* Intricate geometries can be easily made using JCAD3/Takumi and LC-510 in a single run.

The weaknesses of the product and process are as follows:

(1) *Requires support structures.* Structures that have overhangs and undercuts must have supports that are designed and fabricated together with the main structure.

(2) *Requires post-processing.* Post-processing includes removal of supports and other unwanted materials, which is tedious, time-consuming and can damage the model.

(3) *Requires post-curing.* Post-curing may be needed to cure the object completely and ensure the integrity of the structure.

3.7.3   Process

The process building models by the system MEIKO is illustrated in Figure 3.24.

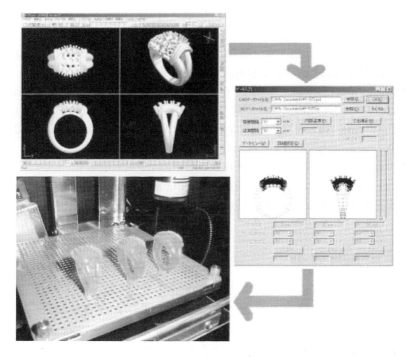

Figure 3.24: Modeling jewelry on JCAD3/Takumi and MEIKO
(Courtesy Meiko Co. Ltd.)

(1) *Designing a model using the exclusive three-dimensional CAD software*: A model is created in the personal computer using JCAD3/Takumi, specifically created for jewelry design in Meiko.

(2) *Creating NC data from the CAD data directly by the exclusive CAM module*: From the CAD data, NC data is generated using the CAM module. This is based on the standard software and methods used in CNC machines.

(3) *Forwarding the NC data to the LC-510 machine*: The NC data and codes are transferred to the NC controller in the LC-510 machine via RS-232C cable. This transfer is similar to those used in CNC machines.

(4) *Building the model by the LC-510*: From the down-loaded NC data, the LC-510 optical modeling system will create the jewelry prototype using the laser and the photo-curable resin layer by layer.

3.7.4   Principle

The fundamental principle behind the method is the laser solidification process of photo-curable resins. As with other liquid-based systems, its principles are similar to that described in Section 3.1.4. The main difference is in the controller of the scanning system. The system MEIKO uses an X–Y (plotter) system with an NC controller instead of the galvanometer mirror scanning system.

Parameters which influence the performance and functionality of the system are the properties of resin, the diameter of the beam spot, and the XY resolution of the machine.

3.7.5   Applications

The general application areas of the system MEIKO are as follows:

(1)   Production of jewelry.
(2)   Production of other small models (such as hearing-aids and a part of spectacles frame).

Users of the system MEIKO produce resin models using exclusively the resin UVM-8001 as its material as shown in Figure 3.25.

There are two ways to create actual products from these resin models as follows:

(1)   In the traditional way (see Figure 3.26), rubber models are made, and wax models are fabricated (see Figure 3.27). Then using the lost wax method, actual products can be manufactured (see Figure 3.28).
(2)   These resin models are used for casting, the same as wax models.

3.7.6   Research and Development

Meiko Co. Ltd. researches on the other new machine (including the resin) and the new edition of the JCAD3/Takumi, cooperating with Professor S. Furukawa at Yamanashi University. Meiko will invest more manpower to produce more machines in order to widen its market.

Figure 3.25: Resin models made by the system MEIKO (Courtesy Meiko Co. Ltd.)

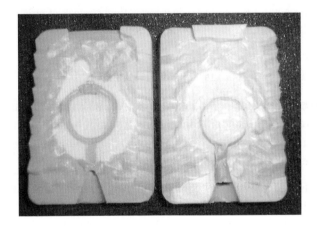

Figure 3.26: Making rubber mold from resin model rings (Courtesy Meiko Co. Ltd.)

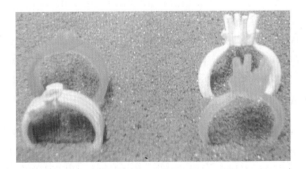

Figure 3.27: Resin ring models and metal casted ring models (Courtesy Meiko Co. Ltd.)

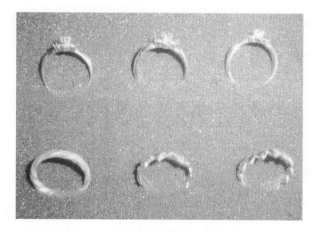

Figure 3.28: Actual precious metal rings casted from originals of resin ring models
(Courtesy Meiko Co. Ltd.)

3.8 OTHER SIMILAR COMMERCIAL RP SYSTEMS

They are several other commercial methods which are similar to what have been described so far. Since they are largely similar in principle, operation and applications, they are not dealt with in details. These systems are summarized as follows:

(1) *SLP.* The manufacturer is Denken Engineering Co. Ltd., 2-1-40, Sekiden-machi Oita City, Oita Prefecture, 870 Japan.

(2) *COLAMM (Computer-Operated, Laser-Active Modeling Machine).* The manufacturer is Mitsui Zosen Corporation of Japan. Opposed to most methods where parts are built on a descending platform, the COLAMM is built on an ascending platform [20]. The first (top) layer of the part is scanned by a laser placed below, through a transparent window plate. The layer is built on the platform and since it is upside down, the part is suspended. The platform is next raised by the programmed layer thickness to allow a new layer of resin to flow onto the window for scanning.

(3) *LMS (Layer Modeling System).* The manufacturer is Fockele und Schwarze Stereolithographietechnik GmbH (F&S) of Borchen-Alfen and its address is Alter Kirchweg 34, W-4799 Borchen-Alfen, Germany. A unique feature of the LMS is its low-cost,

non-optical resin level control system (LCS). The LCS does not contact the resin and therefore allows vats to be changed without system recalibration.

(4) *Light Sculpting.* Not sold commercially, the Light Sculpting device is offered as a bureau service by Light Sculpting of Milwaukee, Wisconsin, USA [20]. The Light Sculpting technique uses a descending platform like SLA and many others, and irradiates the resin surface with a masked lamp as in Cubital's SGC. However, the unique feature of this technique is that the resin is cured in contact with a plate of transparent material on which the mask rests. The resulting close proximity of the mask to the resin surface has good potential of ensuring accurate replication of high-resolution patterns.

3.9 TWO LASER BEAMS

Unlike many of the earlier methods which use a single laser beam, or Cubital's Solid Ground Curing method which uses a lamp, another liquid-based RP method is to employ two laser beams. First conceptualized by Wyn Kelly Swainson in 1967, the method is to penetrate a vat of photocurable resin with two lasers [20]. Opposed to SLA and other similar methods, this method no longer works only on the top surface of a vat of resin. Instead, the work volume resides anywhere in the vat.

In the two-laser-beam method, the two lasers are focused to intersect at a particular point in the vat. The principle behind this method is a two-step variation of ordinary photopolymerization: a single photon initiates curing and two photons of different frequencies are required to initiate polymerization. Therefore, the point of intersection of the two lasers is the desired point of curing. The numerous points of intersection will collectively form the three-dimensional part.

In the method, it is assumed that one laser excites molecules all along its path in the vat. While most of these molecules will return back to their unexcited state after a short time, those also in the path of the second laser will be further excited to initiate the polymerization process. On this basis, the resin cures at the point of intersection of the two laser beams, while the rest lying along either of the two paths will remain liquid.

Unfortunately, after nearly three decades of work on this method, the research work has not materialized into a commercial system, despite the continuation of Swainson's work by Formigraphic Engine Co. in collaboration with Battell Development Corporation, USA, and another independent research group, French National Center for Scientific Research (CNRS), France.

Three major problems need to be resolved before the process is deemed to deliver parts of usable size and resolution. The first two problems relate to the control of the laser beams so that they intersect precisely at the desired three-dimensional point in the vat. First, the focusing problem exists because each laser beam undergoes numerous refraction as its ray moves from the source in air and through the liquid resin. Nonuniformities in the refractive index of the resin are difficult to control and monitor because of the following reasons:

(1) Different progress of curing can cause nonuniformities.
(2) Inherent nonuniformities in the resin itself.
(3) Change of temperature in the vat.

Assuming that the resin can be completely homogenous and temperature changes are minimized so that its effects are insignificant, the second problem to contend with, is how can two laser beams moving at high speed be focused to a very small and accurate point of intersection. The problem is compounded by the high energies required of the two lasers, which means that the two lasers will have short focal lengths. Thus, the distance is limited and in turn, this limits the size of the part that can be made by this method.

Third, it is not always the case that molecules lying along the paths of either laser beams will return to the liquid state. This means unwanted resins may be formed along the path of either laser beams. Consequently, the cured resins may pose further problems to the rest of the process.

3.10   RAPID FREEZE PROTOTYPING

3.10.1   Introduction

Most of the existing rapid prototyping processes are still quite expensive and many of them generate substances such as smoke, dust, hazardous chemicals, etc., which are harmful to human health and the environment. Continuing innovation is essential in order to create new rapid prototyping processes that are fast, clean, and of low-cost.

Dr. Ming Leu of the University of Missouri-Rolla, Virtual & Rapid Prototyping Lab. is developing a novel, environmentally benign rapid prototyping process that uses cheap and clean materials and can achieve good layer binding strength, fine build resolution, and fast build speed. They have invented such a process, called Rapid Freeze Prototyping (RFP), that can make three-dimensional ice parts of arbitrary geometry layer-by-layer by freezing of water droplets.

3.10.2   Objectives

The fundamental study of this process has three objectives:

(1) Generating a good understanding of the physics of this process, including the heat transfer and flow behavior of the deposited material in forming an ice part.
(2) Developing a part building strategy to minimize part build time, while maintaining the quality and stability of the build process.
(3) Investigating the possibility of investment casting application using the ice parts generated by this process.

3.10.3   Advantages and Disadvantages

RFP has the following advantages:

(1) *Low running cost.* Rapid Freeze Prototyping (RFP) process is cheaper and cleaner than all the other rapid prototyping processes. The energy utilization of RFP is low compared with other rapid prototyping processes such as laser stereolithography or selective laser sintering.

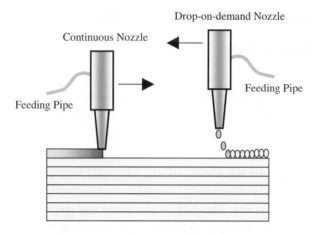

Figure 3.29: Two material extrusion method

(2) *Good accuracy.* RFP can build accurate ice parts with excellent surface finish. It is easy to remove the RFP made ice part in a mold making process, by simply heating the mold to melt the ice part.

(3) *Good building speed.* The build speed of RFP can be significantly faster than other rapid prototyping processes, because a part can be built by first depositing water droplets to generate the part boundary and then filling in the enclosed interior with a water stream (see Figure 3.29). This is possible due to the low viscosity of water. It is easy to build color and transparent parts with the RFP process.

On the other hand, RFP has also the following disadvantages:

(1) *Requires a cold environment.* The prototype of RFP is made of ice and hence it cannot maintain its original shape and form in room temperature.

(2) *Need additional processing.* The prototype made with RFP cannot be used directly but have to be subsequently cast into a mold and so on and this increases the production cost and time.

(3) *Repeatability.* Due to the nature of water, the part built in one run may differ from the next one. The composition of water is also hard to control and determined unless tests are carried out.

3.10.4   Process

An experimental rapid freeze prototyping system for building three-dimensional ice parts has been developed. The experimental system (see Figure 3.30) consists of the following: (a) a three-dimensional positioning subsystem; (b) a material depositing subsystem; and (c) a freezing chamber (see Figure 3.31). This system is being used together

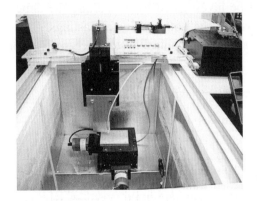

Figure 3.30: Experimental system

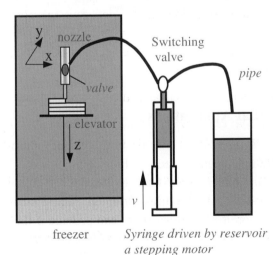

Figure 3.31: Building environment and water extrusion subsystem

with modeling and analysis efforts to understand the behavior of material solidification and fluid spreading during the ice part building process. The process parameters being examined include nozzle scanning speed, environment temperature, fluid density and fluid viscosity. Strategies for building the ice part and its support structure will be developed, with the objective of minimizing the time needed to build the part and its support structure while maintaining process quality and stability. Using the built ice parts, the investment casting application will be investigated by examining the dimensional accuracy and surface finish of the obtained metal castings.

Figure 3.32 shows the control system schematics for the system.

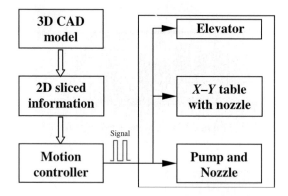

Figure 3.32: Control system schematics

3.10.5   Potential Applications

(1) *Part Visualization.* Parts can be built for the purpose of visualization. Examples can be seen in Figures 3.33 and 3.34.
(2) *Ice Sculpture Fabrication.* One application of rapid freeze prototyping is making ice sculptures for entertainment purposes. Imagine how much more fun a dinner party can provide if there are colorful ice sculptures that can be prescribed one day before and that vary from table to table. This is not an unlikely scenario because RFP can potentially achieve a very fast build speed by depositing water

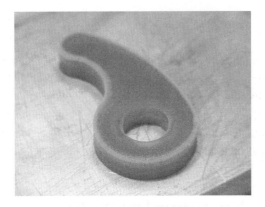

Figure 3.33: Solid link rod (10 mm height)

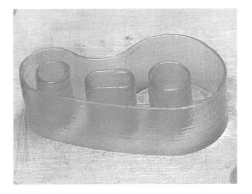

Figure 3.34: Contour of a link rod

droplets only for the part boundary and filling in the interior with a water stream in the ice sculpture fabrication process. Figure 3.35 shows the CAD part, while Figure 3.36 shows the part made by RFP.

(3) *Silicon Molding.* The experiments on UV silicone molding have shown that it is feasible to make silicone molds with ice patterns (see Figures 3.37–3.41). The key advantage of using ice patterns instead of plastic or wax patterns is that ice patterns are easier to remove (without pattern expansion) and no demolding step is needed before injecting urethane or plastic parts. This property can

Figure 3.35: CAD part

Figure 3.36: Part made by RFP

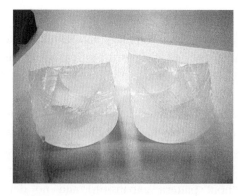

Figure 3.37: UV silicone mold made by ice pattern

Figure 3.38: Urethane part made by UV silicone mold

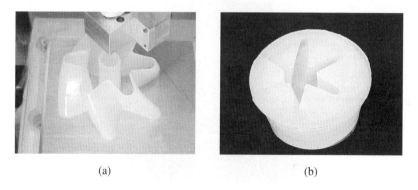

(a)   (b)

Figure 3.39: UV silicone mold made by ice part

Figure 3.40: Urethane part made by UV silicone mold

Figure 3.41: Metal part made by UV silicone mold

Figure 3.42: Metal model made by investment casting with ice pattern

avoid demolding accuracy loss and can allow more complex molds to be made without the time-consuming and experience-dependent design of demolding lines.

(4) *Investment Casting.* A promising industrial application of the RFP technology is investment casting. DURAMAX company recently developed the Freeze Cast Process (FCP), a technology of investment casting with ice patterns made by molding. The company has demonstrated several advantages of this process over the competing wax investment and other casting processes, including low cost (35%–65% reduction), high quality, fine surface finish, no shell

cracking, easy process operation, and faster run cycles. Additionally, there is no smoke and smells in investment casting with ice patterns. Figure 3.42 shows a metal model made by investment casting using ice patterns from RFP.

3.11 MICROFABRICATION

In microfabrication, many methods have been developed and improved over the last few years whereby miniature objects are built. For example, Koji Ikuta and his group in Kyushu Institute of Technology in Japan worked on a process of microfabrication called "integrated hardened polymer stereolithography" (IH process) to build prototype venous valve and three-dimensional micro integrated fluid system (MIFS) [21, 22]. These are very small structures. For instance, the valve's inner diameter is 80 microns (a micron is 0.001 millimeter or about 40 millionths of an inch). Using the IH process, they have also built a functioning micro-electrostatic actuator approximately 700 microns tall and 120 microns wide, consisting of two adjacent bars. When charged with electricity, these bars close a 50-micron gap. Such a switch is used for small mechanical devices such as the micro-venous valve.

Recently, they improved the process, naming it "Super IH process" [23]. The unique feature of this process is that liquid UV (Ultra Violet) polymer can be solidified at a pinpoint position in 3D space by optimizing the apparatus and focusing the laser beam. This pinpoint exposure allows the 3D micro structure to be made without any supporting parts or sacrificial layers. Hence, freely movable micro mechanisms such as gear rotators and free connecting chains can be made easily. Since the total fabrication time of the super IH process is extremely fast, it is easily applied to mass production of 3D microstructures.

Similar micro lithography works have been observed in the prototyping of multidirectional inclined structures [24], flexible microactuator (FMA) [25], and piezoelectric micropump and microchannels [26].

Another method known as the LIGA technique is a new method for microstructure fabrication [27]. LIGA (Lithographie, Galvanoformung und Abformung) is a German abbreviation as it was developed in Germany and it means lithography, galvanoforming and plastic molding.

Synchrotron radiation X-ray lithography is used due to its good parallelism, high radiation intensity and broad spectral range. The template having a thickness of several hundred microns can be fabricated with aspect ratio up to 100 and a two-dimensional structure with submicron deviation is obtained. This structure can then be transformed into a metal or plastic microstructural product by galvanoforming and plastic molding.

The LIGA technique has been successfully used in making engines, turbines, pumps, electric motors, gears, connectors and valves [27]. In the construction of milliactuators, Lehr explained that the LIGA technique offers a large variety of materials to fabricate components with submicron tolerances so as to meet specific actuator functions [28].

X-ray LIGA relies on synchrotron radiation to obtain necessary X-ray fluxes and uses X-ray proximity printing [29]. It is being used in micromechanics (micromotors, microsensors, spinnerets, etc.), micro-optics, micro-hydrodynamics (fluidic devices), microbiology, medicine, biology, and chemistry for microchemical reactors. It compares to micro-electromechanical systems (MEMS) technology, offering a larger, non-silicon choice of materials and better inherent precision. Inherent advantages are its extreme precision, depth of field and very low intrinsic surface roughness. However, the quality of fabricated structures often depends on secondary effects during exposure and effects like resist adhesion.

Deep X-ray lithography (DXRL) is able to create tall microstructures with heights ranging from 100 μm to 1000 μm and aspect ratios of 10 to 50 [30]. DXRL provides lithographic precision of placement, small feature sizes in the micrometer range and sub-micrometer details over the entire height of the structures. Figure 3.43 shows a SEM picture of a gear train assembly.

UV-LIGA relies on thick UV resists as an alternative for projects requiring less precision [29]. Modulating the spectral properties of synchrotron radiation, different regimes of X-ray lithography lead to (a) the mass-fabrication of classical nanostructures, (b) the fabrication of high aspect ratio nanostructures (HARNST), (c) the fabrication of high aspect ratio microstructures (HARMST), and (d) the fabrication of high aspect ratio centimeter structures (HARCST).

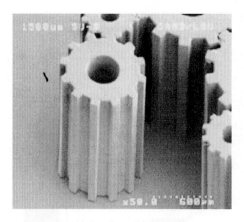

Figure 3.43: SEM picture of a gear train assembly (Courtesy Louisiana State University)

Microstereolithography (MSL) is very similar to stereolithography. The manufacture of 3D micro-objects by using a SL technique first needs a strong correction in the process control, in order to have an accuracy of less than 10 μm in the three directions of space [31]. This method differs from the SL method in that the focus point of the laser beam remains fixed on the surface of the resin, while an *x–y* positioning stage moves the resin reactor in which the object is made. However, the reactor must be translated very slowly to ensure that the surface of the liquid resin is stable during polymerization. As such, the outer size of the microstructure has to be limited unless a long manufacturing time is allowed. The fabrication of 20 μm thick ceramic micro-components has been achieved with this method.

The apparatus consists of a He–Cd laser with acoustic-optic shutter controlled by the computer as shown in Figure 3.44 [32]. The laser beam is then deflected by two computer-controlled low inertia galvanometric mirrors with the aid of focusing lens onto the open surface of the polymer containing photoinitiators. An *XYZ* positioner moves the reactor containing the polymer and the laser beam is focused on the layer to be solidified.

Multifunctional smart materials involve the integration of polymers and nanoceramic particles by chemical bonding as side groups on a polymer backbone. The concept is to design a backbone with functional

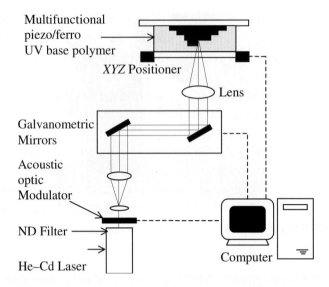

Figure 3.44: Schematic diagram of the Microstereolithography unit

groups that will serve as anchor points for the metal oxides. The nanoparticles such as PZT, PLZT, etc. must have active surfaces or functional groups that can bond with the polymer chain. The nanoparticles provide the piezoelectric function in the polymer and the backbone provides mechanical strength and structural integrity, electrical conductivity, etc. The multifunctionality of these polymers provides a large-scale strain under electric field and thus can be used as actuators for MEMS based devices such as micro pumps.

Functional and structural ceramic materials possess unique properties such as high temperature/chemical resistance, low thermal conductivity, ferroelectricity and piezoelectricity, etc. 3D ceramic microstructures are of special interst in applications such as micro-engines and micro-fluidics. The fabrication of ceramic microstructures differs from that of polymeric MSL. In ceramic MSL, the homogeneous ceramic suspension is prepared. Submicron ceramic powders are mixed with monomer, photoinitiator, dispersant, dilutents, etc. by ball milling for several hours. The prepared ceramic suspension is then put into the vat and ready for MSL based on the CAD design. After MSL, the green body ceramic micro-parts are then obtained. To obtain the dense micro ceramic parts, the green body is next put into a furnace to burn out the polymer binders and further sintered in a high temperature furnace. The binder burnout and the sintering temperature vary with different polymers and ceramics. After sintering, ceramic microstructures are ready for assembly and application.

MSL can be very useful for building microparts in micromechanics, microbiotics (microactuators) and microfluidics [31]. Current lithographic processes previously mentioned have the limitation that complex structures cannot be made easily. Thus, MSL can be used for more complex geometries.

REFERENCES

[1] Rapid Prototyping Report. *3D Systems Introduces Upgraded SLA-250 with Zephyr Recoating* **6**(4) (April 1996): 3.

[2] 3D Systems Product brochure. *SLA Series*, 1999.

[3] Jacobs, P.F., *Rapid Prototyping & Manufacturing, Fundamentals of Stereolithography*, Society of Manufacturing Engineers, 1992, Chapter 1: 11–18.

[4] Wilson, J., *Radiation Chemistry of Monomers, Polymers, and Plastics*, Marcel Dekker, NY, 1974.

[5] Lawson, K., "UV/EB curing in North America," *Proceedings of the International UV/EB Processing Conference*, Vol. 1, May 1–5, 1994, Florida, USA.

[6] Reiser, A., *Photosensitive Polymers*, John Wiley, NY, 1989.

[7] Jacobs, P.F., *Rapid Prototyping & Manufacturing, Fundamentals of Stereolithography*, Society of Manufacturing Engineers, 1992, Chapter 2: 25–32.

[8] Jacobs, P.F., *Stereolithography and other RP&M Technologies*, Society of Manufacturing Engineers, 1996, Chapter 2: 29–35.

[9]   Jacobs, P.F., *Rapid Prototyping & Manufacturing, Fundamentals of Stereolithography*, Society of Manufacturing Engineers, 1992, Chapter 2: 53–56.

[10]   Jacobs, P.F., *Rapid Prototyping & Manufacturing, Fundamentals of Stereolithography*, Society of Manufacturing Engineers, 1992, Chapter 3: 60–78.

[11]   Rapid Prototyping Report, *Ford Uses Stereolithography to Cast Production Tooling* **4**(7) (July 1994): 1–3.

[12]   Rapid Prototyping Report, *Ciba Introduces Fast Polyurethanes for Part Duplication* **6**(3) (March 1996): 6.

[13]   Jacobs, P.F., "Insight: Moving toward rapid tooling," *The Edge* **IV**(3) (1995): 6–7.

[14]   Kobe, G., "Cubital's unknown solider," *Automotive Industries* (August 1992): 54–55.

[15]   Johnson, J.L., *Principles of Computer Automated Fabrication*, Palatino Press, 1994, Chapter 2: 44.

[16]   Cubital News Release, *Cubital Prototyping Machine Builds Jeep in 24 Hours*, July, 1995.

[17]   Cubital News Release, *New Cubital-based SoliCast Process Offers Inexpensive Metal Prototypes in Two Weeks*, April, 1995.

[18]   Cubital New Release, *Cubital Saves The Day — In Fact, About Six Weeks — For Pump Developers*, March, 1995.

[19]   CMET Inc. Product Brochure, *SOUPII 600GS Series*, 2002.

[20]   Burns, M., *Automated Fabrication: Improving Productivity in Manufacturing*, PTR Prentice Hall, 1993.

[21]   Rapid Prototyping Report, *Microfabrication* **4**(7) (July 1994): 4–5.

[22]   Ikuta, K., Hirowatari, K., and Ogata, T., "Three-dimensional micro integrated fluid systems (MIFS) fabricated by stereolithography,"

Proceedings of the IEEE Micro Electro Mechanical Systems, Oiso, Japan, 1994.

[23] Ikuta, K., Maruo, S., and Kojima, S., "New micro stereo lithography for freely movable 3D micro structure — Super IH process with submicron resolution," *Proceedings of the IEEE Micro Electro Mechanical Systems*, 1998, Piscataway, NJ, USA: 290–295.

[24] Beuret, C., Racine, G.A., Gobet, J., Luthier, R., and de Rooij, N.F., "Microfabrication of 3D multidirectional inclined structures by UV lithography and electroplating," *Proceedings of the IEEE Micro Electro Mechanical Systems*, Oiso, Japan, 1994.

[25] Suzomori, K., Koga, A., and Haneda, R., "Microfabrication of integrated FMAs using stereo lithography," *Proceedings of the IEEE Micro Electro Mechanical Systems*, Oiso, Japan, 1994.

[26] Carrozza, M.C., Croce, N., Magnani, B., and Dario, P., "Piezoelectric-drive stereolithography-fabricated micropump," *Journal of Micromechanics and Microengineering* **5**(2) (June 1995): 177–179.

[27] Yi, F., Wu, J., and Xian, D., "LIGA technique for microstructure fabrication," *Weixi Jiagong Jishu/Microfabrication Technology*, No. 4 (December 1993): 1–7.

[28] Lehr, H., Ehrfeld, W., Kaemper, K.P., Michel, F., and Schmidt, M., "LIGA components for the construction of milliactuators," *Proceedings of the 1994 IEEE Symposium on Emerging Tehnologies and Factory Automation*, Tokyo, Japan: 1993, 43–47.

[29] Kupka, R.K., Bouamrane, F., Cremers, C., and Megtert, S., "Microfabrication: LIGA-X and applications," *Applied Surface Science* **164** (September 2000): 97–110.

[30] Aigeldinger, G., Coane, P., Braft, B., Goettert, J., Ledger, S., Zhong, G.L., Manohara, H., and Rupp, L., "Preliminary results at the Ultra Deep X-ray lithography beamline at CAMD,"

Proceedings of SPIE Vol. 4019, Design, Test, Integration, and Packaging of MEMS/MOEMS, Paris, France: 2000, 429–435.

[31] Basrour, S., Majjad, H., Coudevylle, J.R., and de Labachelerie, M., "Complex ceramic-polymer composite microparts made by microstereolithography," *Proceedings of SPIE Vol. 4408, Design, Test, Integration, and Packaging of MEMS/MOEMS*, Cannes, France: 2001, 535–542.

[32] Varadan, V.K. and Varadan, V.V., "Micro stereo lithography for fabrication of 3D polymeric and ceramic MEMS," *Proceedings of SPIE Vol. 4407, MEMS Design, Fabrication, Characterization, and Packaging*, Edinburgh, UK: 2001, 147–157.

PROBLEMS

1. Describe the process flow of the 3D System Stereolithography Apparatus.

2. Describe the process flow of Cubital's Solid Ground Curing System.

3. Compare and contrast the laser-based stereolithography systems and the solid ground curing systems. What are the advantages (and disadvantages) for each of the systems.

4. Describe how investment casting parts can be made using:
   (i)   3D Systems' SLA
   (ii) Cubital's SGC system
   (iii) CMET's SOUP system and
   (iv) Teiji Seiki's SOLIFORM

5. Which liquid-based machine has the largest work volume? Which has the smallest?

6. Meiko Co. Ltd. produces the LC-315 for jewelry prototyping. By comparing the machine specifications with other vendors, discuss what you think are the important specifications that will determine their suitability for jewelry prototyping.

7. As opposed to many of the liquid-based RP systems which uses photosensitive polymer, water is used in the Rapid Freeze Prototyping (RFP). What are the pros and cons of using water?

8. Discuss the principle behind the two-laser-beam method. What are the major problems in this method?

Chapter 4
SOLID-BASED RAPID PROTOTYPING SYSTEMS

Solid-based rapid prototyping systems are very different from the liquid-based photo-curing systems described in Chapter 3. They are also different from one another, though some of them do use the laser in the prototyping process. The basic common feature among these systems is that they all utilize solids (in one form or another) as the primary medium to create the prototype. A special group of solid-based RP systems that uses powder as the prototyping medium will be covered separately in Chapter 5.

4.1 CUBIC TECHNOLOGIES' LAMINATED OBJECT MANUFACTURING (LOM™)

4.1.1 Company

Cubic Technologies was established in December 2000 by Michael Feygin, the inventor who developed Laminated Object Manufacturing® (LOM™). In 1985, Feygin set up the original company, Helisys Inc., to market the LOM™ rapid prototyping machines. However, sales figures did not meet up to expectations [1] and the company ran into financial difficulties. Helisys Inc. subsequently ceased operation in November 2000. Currently, Cubic Technologies, the successor to Helisys Inc., is the exclusive manufacturer of the LOM™ rapid prototyping machine. The company's address is Cubic Technologies Inc., 100E, Domingnez Streets, Carson, California 90746-3608, USA.

4.1.2   Products

4.1.2.1   *Models and Specifications*

Cubic Technologies offers two models of LOM™ rapid prototyping systems, the LOM-1015Plus™ and LOM-2030H™ (see Figure 4.1). Both these systems use the CO_2 laser, with the LOM-1015Plus™ operating a 25 W laser and the LOM-2030H™ operating a 50 W laser. The optical system, which delivers a laser beam to the top surface of the work, consists of three mirrors that reflect the CO_2 laser beam and a focal lens that focuses the laser beam to about 0.25 mm (0.010"). The control of the laser during cutting is by means of a *XY* positioning table that is servo-based as opposed to the galvanometer mirror system. The LOM-2030H™ is a larger machine and produces larger prototypes. The work volume of the LOM-2030H™ is 810 mm × 550 mm × 500 mm (32" × 22" × 20") and that of the LOM-1015Plus™ is 380 mm × 250 mm × 350 mm (15" × 10" × 14"). Detailed specifications of the two machines are summarized in Table 4.1.

4.1.3   Process

The patented Laminated Object Manufacturing® (LOM™) process [2–4] is an automated fabrication method in which a 3D object is constructed from a solid CAD representation by sequentially laminating the part

Figure 4.1: The LOM-2030H™ (left) and the LOM-1015Plus™ (right) (Courtesy Cubic Technologies Inc.)

Table 4.1: Specifications of LOM-1015Plus™ and LOM-2030H™

Model	LOM-1015Plus™	LOM-2030H™
Max. part envelope size, mm (in)	L381 × W254 × H356 (L15 × W10 × H14)	L813 × W559 × H508 (L32 × W22 × H20)
Max. part weight, kg (lbs)	32 (70)	204 (405)
Laser, power and type	Sealed 25 W, CO_2 Laser	Sealed 50 W, CO_2 Laser
Laser beam diameter, mm (in)	0.20–0.25 (0.008–0.010)	0.203–0.254 (0.008–0.010)
Motion control	Servo-based X–Y motion systems with a speed up to 457 mm/sec (18"/sec); Typical Z-platform feedback for motion system	Brushless servo-based X–Y motion dystems with a speed up to 457 mm/sec (18"/sec); Typical Z-platform feedback for motion system
Part accuracy XYZ directions, mm (in)	±0.127 mm (±0.005 in)	±0.127 mm (±0.005 in)
Material thickness, mm (in)	0.08–0.25, (0.003–0.008)	0.076–0.254, (0.003–0.008)
Material size	Up to 356 mm (14") roll width and roll diameter	Up to 711 mm (28") roll width and roll diameter
Floor space, m (ft)	3.66 × 3.66 (12 × 12)	4.88 × 3.66 (16 × 12)
Power	Two (2) 110VAC, 50/60 Hz, 20 Amp, single phase Two (2) 220VAC, 50/60 Hz, 15 Amp, single phase	220VAC, 50/60 Hz, 30 Amp, single phase
Materials	LOMPaper® LPH series, LPS series LOMPlastics® LPX series	LOMPaper® LPH series, LPS series LOMPlastics® LPX series, LOMComposite® LGF series

cross-sections. The process consists of three phases: pre-processing; building; post-processing.

4.1.3.1 *Pre-processing*

The pre-processing phase comprises several operations. The initial steps include generating an image from a CAD-derived STL file of the part to be manufactured, sorting input data, and creating secondary data structures. These are fully automated by LOMSlice™, the LOM™ system software, which calculates and controls the slicing functions. Orienting and merging the part on the LOM™ system are done manually. These tasks are aided by LOMSlice™, which provides a menu-driven interface to perform transformations (e.g., translation, scaling, and mirroring) as well as merges.

4.1.3.2 *Building*

In the building phase, thin layers of adhesive-coated material are sequentially bonded to each other and individually cut by a CO_2 laser beam (see Figure 4.2). The build cycle has the following steps:

(1) LOMSlice™ creates a cross-section of the 3D model measuring the exact height of the model and slices the horizontal plane accordingly. The software then images crosshatches which define the outer perimeter and convert these excess materials into a support structure.

(2) The computer generates precise calculations, which guide the focused laser beam to cut the cross-sectional outline, the cross-hatches, and the model's perimeter. The laser beam power is designed to cut exactly the thickness of one layer of material at a time. After the perimeter is burned, everything within the model's boundary is "freed" from the remaining sheet.

(3) The platform with the stack of previously formed layers descends and a new section of material advances. The platform ascends and the heated roller laminates the material to the stack with a single reciprocal motion, thereby bonding it to the previous layer.

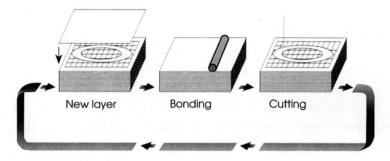

Figure 4.2: LOM™ building process (Courtesy Cubic Technologies Inc.)

(4) The vertical encoder measures the height of the stack and relays the new height to LOMSlice™, which calculates the cross section for the next layer as the laser cuts the model's current layer.

This sequence continues until all the layers are built. The product emerges from the LOM™ machine as a completely enclosed rectangular block containing the part.

4.1.3.3 *Post-processing*

The last phase, post-processing, includes separating the part from its support material and finishing it. The separation sequence is as follows [see Figures 4.3(a)–4.3(d)]:

(1) The metal platform, home to the newly created part, is removed from the LOM™ machine. A forklift may be needed to remove the larger and heavier parts from the LOM-2030H™.
(2) Normally a hammer and a putty knife are all that is required to separate the LOM™ block from the platform. However, a live thin wire may also be used to slice through the double-sided foam tape, which serves as the connecting point between the LOM™ stack and the platform.
(3) The surrounding wall frame is lifted off the block to expose the crosshatched pieces of the excess material. Crosshatched pieces may then be separated from the part using wood carving tools.

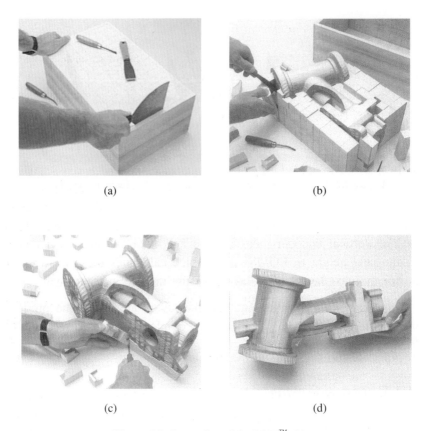

Figure 4.3: Separation of the LOM™ object
(a) The laminated stack is removed from the machine's elevator plate.
(b) The surrounding wall is lifted off the object to expose cubes of excess material.
(c) Cubes are easily separated from the object's surface.
(d) The object's surface can then be sanded, polished or painted, as desired.

After the part is extracted from surrounding crosshatches the wood-like LOM™ part can be finished. Traditional model-making finishing techniques, such as sanding, polishing, painting, etc. can be applied. After the part has been separated it is recommended that it be sealed immediately with urethane, epoxy, or silicon spray to prevent moisture absorption and expansion of the part. If necessary, LOM™ parts can be machined — by drilling, milling and turning.

4.1.3.4 *System Structure*

The LOM-1015Plus™ and LOM-2030H™ have a similar system structure which can be broken down into several subsystems: computer hardware and software, laser and optics, $X-Y$ positioning device, platform and vertical elevator, laminating system, material supply and take-up.

The computer is an IBM-compatible PC. The LOM™ software, LOMSlice™, is a true 32-bit application with a user-friendly interface including menus, dialog boxes and progress indicators. LOMSlice™ is completely integrated, providing preprocessing, slicing, and machine control within a single program. Z-dimension accuracy is maintained through a closed loop real-time feedback mechanism and is calculated upon each lamination. As the laser is cutting the model, software is simultaneously planning the next layer's outline and crosshatches.

LOMSlice™ can also overcome STL file imperfections that violate facets of normal vector orientation or vertex-to-vertex rules [5], or even those missing facets. In order to facilitate separation of the part from excess material, LOMSlice™ automatically assigns (or "burns out") reduced crosshatch sizes to intricate regions.

To make it easier, faster and safer to align the laser beam, a helium-neon visible laser which projects a red beam of light and is collinear with the live CO_2 laser beam is used. The operator can switch on the innocuous red laser beam and watch as the mirrors are aligned, rather than using trial and error with the invisible and powerful CO_2 beam.

Lamination is accomplished by applying heat and pressure by way of rolling a heated cylinder across the sheet of material, which has a thin layer of a thermoplastic adhesive on one side. Studies [6] have indicated that interlaminate strength of LOM™ parts is a complex function of bonding speed, sheet deformation, roller temperature, and contact area between the paper and the roller. By increasing pressure of the heated roller, lamination is improved due to fewer air bubbles. Increased pressure also augments the contact area thereby bolstering interlaminate strength. Pressure is controlled by the limit switch which is mounted on the heated roller. If compression is set too high it can cause distortion in the part.

The material supply and take-up system comprises two material roll supports (supply and rewind), several idle rollers to direct the material, and two rubber-coated nip-rollers (driving and idle), which advance or rewind the sheet material during the preprocessing and building phases. To make material flow through the LOM™ systems more smoothly, mechanical nip rollers are used. The friction resulting from compressing moving material between the rubber coated roller on both the feed and wind mechanism ensures a clean feed and avoids jamming.

4.1.3.5 *Materials*

Potentially, any sheet material with adhesive backing can be utilized in Laminated Object Manufacturing. It has been demonstrated that plastics, metals, and even ceramic tapes can be used. However, the most popular material has been Kraft paper with a polyethylene-based heat seal adhesive system because it is widely available, cost-effective, and environmentally benign [7].

In order to maintain uniform lamination across the entire working envelope it is critical that the temperature remain constant. A temperature control system, with closed-loop feedback, ensures the system's temperature remains constant, regardless of its surrounding environment.

4.1.4 Principle

The LOM™ process is based on the following principles:

(1) Parts are built, layer-by-layer, by laminating each layer of paper or other sheet-form materials and the contour of the part on that layer is cut by a CO_2 laser.

(2) Each layer of the building process contains the cross-sections of one or many parts. The next layer is then laminated and built directly on top of the laser-cut layer.

(3) The Z-control is activated by an elevation platform, which lowers when each layer is completed, and the next layer is then laminated and ready for cutting. The Z-height is then measured for the exact

height so that the corresponding cross sectional data can be calculated for that layer.

(4) No additional support structures are necessary as the "excess" material, which are cross-hatched for later removal, act as the support.

4.1.5   Advantages and Disadvantages

The main advantages of using LOM™ technology are as follows:

(1) *Wide variety of materials.* In principle, any material in sheet form can be used in the LOM™ systems. These include a wide variety of organic and inorganic materials such as paper, plastics, metals, composites and ceramics. Commercial availability of these materials allow users to vary the type and thickness of manufacturing materials to meet their functional requirements and specific applications of the prototype.

(2) *Fast build time.* The laser in the LOM™ process does not scan the entire surface area of each cross-section, rather it only outlines its periphery. Therefore, parts with thick sections are produced just as quickly as those with thin sections, making the LOM™ process especially advantageous for the production of large and bulky parts.

(3) *High precision.* The feature to feature accuracy that can be achieved with LOM™ machines is usually better than 0.127 mm (0.005"). Through design and selection of application specific parameters, higher accuracy levels in the $X-Y$ and Z dimensions can be achieved. If the layer does shrink horizontally during lamination, there is no actual distortion as the contours are cut post-lamination, and laser cutting itself does not cause shrinkage. If the layers shrink in the transverse direction, a closed-loop feedback system gives the true cumulative part height upon each lamination to the software, which then slices the 3D model with a horizontal plane at the appropriate location.

The LOM™ system uses a precise $X-Y$ positioning table to guide the laser beam; it is monitored throughout the build process by the closed-loop, real-time motion control system, resulting in an accuracy of ±0.127 mm regardless of the part size. The Z-axis is also controlled using a real-time, closed-loop feedback system. It measures the cumulative part height at every layer and then slices the CAD geometry at the exact Z location. Also, as the laser cuts only the perimeter of a slice there is no need to translate vector data into raster form, therefore the accuracy of the cutting depends only on the resolution of the CAD model triangulation.

(4) *Support structure.* There is no need for additional support structure as the part is supported by its own material that is outside the periphery of the part built. These are not removed during the LOM™ process and therefore automatically act as supports for its delicate or overhang features.

(5) *Post-curing.* The LOM™ process does not need to convert expensive, and in some cases toxic, liquid polymers to solid plastics or plastic powders into sintered objects. Because sheet materials are not subjected to either physical or chemical phase changes, the finished LOM™ parts do not experience warpage, internal residual stress, or other deformations.

The main disadvantages of using LOM™ are as follows:

(1) *Precise power adjustment.* The power of the laser used for cutting the perimeter (and the crosshatches) of the prototype needs to be precisely controlled so that the laser cuts only the current layer of lamination and not penetrate into the previously cut layers. Poor control of the cutting laser beam may cause distortion to the entire prototype.

(2) *Fabrication of thin walls.* The LOM™ process is not well suited for building parts with delicate thin walls, especially in the Z-direction. This is because such walls usually are not sufficiently rigid to withstand the post-processing process when the cross-hatched outer perimeter portion of the block is being removed. The person performing the post-processing task of separating the thin wall of the part from its support must be fully aware of where such delicate parts are located in the model and take sufficient precautions so as not to damage these parts.

(3) *Integrity of prototypes.* The part built by the LOM™ process is essentially held together by the heat sealed adhesives. The integrity of the part is therefore entirely dependent on the adhesive strength of the glue used, and as such is limited to this strength. Therefore, parts built may not be able to withstand the vigorous mechanical loading that the functional prototypes may require.

(4) *Removal of supports.* The most labor-intensive part of the LOM™ process is its last phase of post-processing when the part has to be separated from its support material within the rectangular block of laminated material. This is usually done with wood carving tools and can be tedious and time consuming. The person working during this phase needs to be careful and aware of the presence of any delicate parts within the model so as not to damage it.

4.1.6 Applications

LOM™'s applicability is across a wide spectrum of industries, including industrial equipment for aerospace or automotive industries, consumer products, and medical devices ranging from instruments to prostheses. LOM™ parts are ideal in design applications where it is important to visualize what the final piece will look like, or to test for form, fit and function; as well as in a manufacturing environment to create prototypes, make production tooling or even produce a small volume of finished goods.

(1) *Visualization.* Many companies utilize LOM™'s ability to produce exact dimensions of a potential product purely for visualization. LOM™ part's wood-like composition allows it to be painted or finished as a true replica of the product. As the LOM™ procedure is inexpensive several models can be created, giving sales and marketing executives opportunities to utilize these prototypes for consumer testing, marketing product introductions, packaging samples, and samples for vendor quotations.

(2) *Form, fit and function.* LOM™ parts lend themselves well for design verification and performance evaluation. In low-stress environments LOM™ parts can withstand basic tests, giving

manufacturers the opportunity to make changes as well as evaluate the aesthetic property of the prototype in its total environment.

(3) *Manufacturing.* The LOM™ part's composition is such that, based on the sealant or finishing products used, it can be further tooled for use as a pattern or mold for most secondary tooling techniques including: investment casting, casting, sanding casting, injection molding, silicon rubber mold, vacuum forming and spray metal molding. LOM™ parts offer several advantages important for the secondary tooling process, namely: predictable level of accuracy across the entire part; stability and resistance to shrinkage, warpage and deformity; and the flexibility to create a master or a mold. In many industries the master created through secondary tooling, or even when the LOM™ part serves as the master (e.g., vacuum forming), withstands enough injections, wax shootings or vacuum pressure to produce a low production run from 5 to 1000 pieces.

(4) *Rapid tooling.* Two part negative tooling is easily created with LOM™ systems. Since the material is solid and inexpensive, bulk complicated tools are cost effective to produce. These wood-like molds can be used for injection of wax, polyurethane, epoxy or other low pressure and low temperature materials. Also, the tooling can be converted to aluminum or steel via the investment casting process for use in high temperature molding processes.

4.1.7   Example

4.1.7.1   *National Aeronautical and Space Administration (NASA) and Boeing Rocketdyne Uses LOM™ to Create Hot Gas Manifold for Space Shuttle Main Engine [8]*

One successful example of how an organization implements LOM™ systems into their design process would be from the Rapid Prototyping Laboratory, NASA's Marshall Space Flight Center (MSFC), Huntville, AL.

The laboratory was set up initially to conduct research and development in different ways to advance the technology of building parts in space by remote processing methods. However, as MSFC engineers found a lot more useful applications, i.e., production of concept models and proof-out of component designs other than remote processing when rapid prototyping machines were installed, the center soon became a rapid prototyping shop for other MSFC groups, as well as other NASA locations and NASA subcontractors.

The center acquired the LOM-1015™ machine from Helisys in 1999 to add on their existing rapid prototyping systems and the machine was put through its first challenge when MSFC's contractor, Boeing/Rocketdyne designed a hot gas manifold for the space shuttle's main engine. The part measured 2.40 m (8 ft) long and 0.10 m (4 in) in diameter and was complex in design with many twists and turns and "tee" junction connectors. If the conventional method of creating the prototype were employed, it would require individual steel parts to be welded together to form the prototype. However, there was always a potential of leakage at the joint part and thus, an alternate method was considered. The prototype was to be made from a single piece of steel and such a solution was not only expensive, the prototype built did not fit well to the main engine of the space shuttle.

Eventually, engineers at Boeing decided to build the part using the LOM™ process at MSFC. They prepared a CAD drawing of the design and sent it over to MSFC. The design was sectioned into eight parts, each with the irregular boss-and-socket built in them so as to facilitate joining of the parts together upon completion.

The whole building process took ten days to complete, including three days of rework for flawed parts. It was worked on continuously. One advantage of using the LOM™ machine is that the system can be left unattended throughout the building process and if the system runs out of paper or the paper gets jammed while building, it is able to alert the operator via a pager. The prototype was then mounted onto the actual space shuttle for final fit check analysis. It was estimated that the company saved tens of thousands of dollars, although Boeing declined to reveal the actual cost saving. The whole process also drastically reduces the building time from two to three months to a mere ten days.

4.2   STRATASYS' FUSED DEPOSITION MODELING (FDM)

4.2.1   Company

Stratasys Inc. was founded in 1989 and has developed most of the company's products based on the Fused Deposition Modeling (FDM) technology. The technology was first developed by Scott Cramp in 1988 and the patent was awarded in the U.S. in 1992. FDM uses the extrusion process to build 3D models. Stratasys introduced its first rapid prototyping machine, the 3D modeler® in early 1992 and started shipping the units later that year. Over the past decade, Stratasys has grown progressively, seeing her rapid prototyping machines' sales increase from six units in the beginning to a total of 1582 units in the year 2000 [9]. The company's address is Stratasys Inc., 14950 Martin Drive, Eden Prairie, MN 55344-202, USA.

4.2.2   Products

4.2.2.1   *Models and Specifications*

Stratasys has developed a series of rapid prototyping machines and also a wide range of modeling materials to cater to various industries' needs. The company's rapid prototyping systems can be broadly classified into two categories, the FDM series and the concept modeler. The FDM series include models like FDM 3000, FDM Maxum and FDM Titan. The concept modeler series includes models like Dimension and Prodigy Plus. A summary of the product specifications is found in Tables 4.2 and 4.3.

4.2.2.2   *FDM Series*

The FDM series provide customers with a comprehensive range of versatile rapid prototyping systems. These high-end systems (see Figure 4.4) are not only able to produce 3D models for mechanical testing, they are also able to produce functional prototypes that work as well as a production unit. For older systems like the FDM 3000,

Table 4.2: Specifications of FDM series (Courtesy of Stratasys Inc.)

Models	FDM 3000	FDM Maxum	FDM Titan
Technology	FDM		
Build size, mm (in)	Parts up to 254 × 254 × 406 (10 × 10 × 16)	Parts up to 600 × 500 × 600 (23 × 19.7 × 23)	Parts up to 355 × 406 × 406 (14 × 16 × 16)
Accuracy, mm (in)	± 0.127 (± 0.005)	Up to 127 mm (5 in): ± 0.127 (± 0.005) Greater than 127 mm (5 in) ± 0.038 mm/mm (± 0.0015 in/in)	
Layer road width, mm (in)	0.250 to 0.965 (0.010 to 0.038)	0.305 to 0.965 (0.012 to 0.038)	Not available
Layer thickness, mm (in)	0.178 to 0.356 (0.007 to 0.014)	0.178 to 0.356 (0.007 to 0.014)	0.25 (0.010)
Support structures	Automatically generated with SupportWorks software; WaterWorks or Break-Away Support System (BASS)	Automatically generated with Insight software; WaterWorks soluble support system	
Size, $w \times h \times d$, mm (in)	660 × 1067 × 914 (26 × 42 × 36)	2235 × 1981 × 1118 (88 × 78 × 44)	1270 × 1981 × 876 (50 × 78 × 34.5)
Weight, kg (lbs)	160 (350)	1134 (2500)	726 (1600)
Power requirements	208–240 VAC, 50/60 Hz, 10 A 110–120 VAC, 60 Hz, 20 A	208–240 VAC, 50/60 Hz, 32 A single phase (min. 50 A dedicated service)	230 V, 50/60 Hz, 3 phase, 16 A/phase (min. 20 A dedicated service)
Modeling materials	ABS (White), ABSi, Investment Casting Wax, Elastomer	ABS (white)	ABS, Polycarbonate, Polyphenyl-sulfone
Software	QuickSlice® and Support-Works™	Insight	Insight

Table 4.3: Specifications of Stratasys' concept modelers (Courtesy Stratasys Inc.)

Model	Dimension	Prodigy Plus
Technology	3D printing base on FDM	FDM
Build size, mm (in)	305 × 203 × 203 (12 × 8 × 8)	203 × 203 × 305 (8 × 8 × 12)
Accuracy, mm (in)	± 0.127 (± 0.005)	± 0.127 (± 0.005)
Layer thickness mm (in)	"Standard" — 0.245 (0.010) "Draft" — 0.33 mm (0.013)	User selectable: "Fine" — 0.178 mm (0.007) "Standard" — 0.245 (0.010) "Draft" — 0.33 mm (0.013)
Automatic operation	Easy to use Catalyst™ software imports STL files and automatically slices the model, creates any necessary support structures and generates build files	Catalyst™ software automatically imports and slices STL files, orients the part, generates soluble support structures (if necessary), and creates the deposition path to build parts
Size, $w \times d \times h$, mm (in)	914 × 686 × 1041 (36 × 27 × 41)	864 × 686 × 1041 (34 × 27 × 41)
Weight, kg (lbs)	136 (300)	128 (282)
Power requirements	220–240 VAC, 50/60 Hz, 6 A or 110–120 VAC, 60 Hz, 12 A	110–120 VAC, 60 Hz, 15 A max. or 220–240 VAC, 50/60 Hz, 7 A max
Materials	ABS plastic in white (standard), blue, yellow, black, red or green. Custom colors available	
Material supply	One autoload cartridge with 950 cu. cm. (58 cu. in.) ABS material and one autoload cartridge with 950 cu. cm. (58 cu. in.) support material	One autoload cartridge with 950 cu. cm. (58 cu. in.) ABS material and one autoload cartridge with 950 cu. cm. (58 cu. in.) soluble support material

Figure 4.4: Stratasys' FDM Titan rapid prototyping system (Courtesy Stratasys Inc.)

Quickslice® and SupportWorks™ preprocessing software are used to run with the systems. However, newer systems like the FDM Maxum and Titan use an improved software, Insight. The newer software increases building speed, improves efficiency and is easier to use than its previous QuickSlice® software [10]. Although both Maxum and Titan have the same achievable accuracy, they differ from each other in terms of build volume, layer thickness and their physical size and weight. An advantageous point for selecting Titan over Maxum is that the former allows users to have a wider selection of materials (ABS, Polycarbonate and Polyphenylsulfone), whereas the latter can only build models using ABS.

4.2.2.3　*Concept Modeler Series*

Stratasys produces two concept modelers, the Dimension and the Prodigy Plus. Dimension (see Figure 4.5) uses a 3D printing technology that is based on FDM, which uses a heated head and pump assembly to deposit model plastics onto the build layers. Dimension as a low cost concept modeler helps designers to evaluate products by a quick 3D print of models and eliminates all products' imperfections in the early design stages.

　　Prodigy Plus (see Figure 4.6) replaces the Prodigy which was developed by Stratasys to "fill the void" between the old Genisys Xs and an older version of the FDM series, the FDM 2000 [11]. The

Figure 4.5: Stratasys' Dimension concept modeler (Courtesy Stratasys Inc.)

Figure 4.6: Stratasys' Prodigy Plus concept modeler (Courtesy of Stratasys Inc.)

systems were designed to be used in a networked office environment and to build the 3D conceptual model from any CAD workstation.

Both Dimension and Prodigy Plus have the same build volume, but the two systems differ from each other in many ways. One of the differences is the support material supply that the systems build models with. Dimension builds support structures with the Break Away Support System (BASS™), whereas Prodigy Plus incorporates the WaterWorks automated support system.

4.2.3 Process

In this patented process [12], a geometric model of a conceptual design is created on a CAD software which uses IGES or STL formatted files. It can then imported into the workstation where it is processed through the QuickSlice® and SupportWork™ propriety software before loading to FDM 3000 or similar systems. For FDM Maxum and Titan, a newer software known as Insight is used. The basic function of Insight is similar to that of QuickSlice® and the only difference is that Insight does not need another software to auto-generate the supports. The function is incorporated into the software itself. Within this software, the CAD file is sliced into horizontal layers after the part is oriented for the optimum build position, and any necessary support structures are automatically detected and generated. The slice thickness can be set manually to anywhere between 0.172 to 0.356 mm (0.005 to 0.014 in) depending on the needs of the models. Tool paths of the build process are then generated which are downloaded to the FDM machine.

The modeling material is in spools — very much like a fishing line. The filament on the spools is fed into an extrusion head and heated to a semi-liquid state. The semi-liquid material is extruded through the head and then deposited in ultra thin layers from the FDM head, one layer at a time. Since the air surrounding the head is maintained at a temperature below the materials' melting point, the exiting material quickly solidifies. Moving on the X–Y plane, the head follows the tool path generated by QuickSlice® or Insight generating the desired layer. When the layer is completed, the head moves on to create the next layer. The horizontal width of the extruded material can vary between 0.250 to 0.965 mm depending on model. This feature, called "road width", can vary from slice to slice. Two modeler materials are dispensed through a dual tip mechanism in the FDM machine. A primary modeler material is used to produce the model geometry and a secondary material, or release material, is used to produce the support structures. The release material forms a bond with the primary modeler material and can be washed away upon completion of the 3D models.

4.2.4   Principle

The principle of the FDM is based on surface chemistry, thermal energy, and layer manufacturing technology. The material in filament (spool) form is melted in a specially designed head, which extrudes on the model. As it is extruded, it is cooled and thus solidifies to form the model. The model is built layer by layer, like the other RP systems. Parameters which affect performance and functionalities of the system are material column strength, material flexural modulus, material viscosity, positioning accuracy, road widths, deposition speed, volumetric flow rate, tip diameter, envelope temperature, and part geometry.

4.2.5   Advantages and Disadvantages

The main advantages of using FDM technology are as follows:

(1) *Fabrication of functional parts*. FDM process is able to fabricate prototypes with materials that are similar to that of the actual molded product. With ABS, it is able to fabricate fully functional parts that have 85% of the strength of the actual molded part. This is especially useful in developing products that require quick prototypes for functional testing.

(2) *Minimal wastage*. The FDM process build parts directly by extruding semi-liquid melt onto the model. Thus only those material needed to build the part and its support are needed, and material wastages are kept to a minimum. There is also little need for cleaning up the model after it has been built.

(3) *Ease of support removal*. With the use of Break Away Support System (BASS) and WaterWorks Soluble Support System, support structures generated during the FDM building process can be easily broken off or simply washed away. This makes it very convenient for users to get to their prototypes very quickly and there is very little or no post-processing necessary.

(4) *Ease of material change*. Build materials, supplied in spool form (or cartridge form in the case of the Dimension or Prodigy Plus), are easy to handle and can be changed readily when the materials

in the system are running low. This keeps the operation of the machine simple and the maintenance relatively easy.

The main disadvantages of using FDM technology are as follows:

(1) *Restricted accuracy*. Parts built with the FDM process usually have restricted accuracy due to the shape of the material used, i.e., the filament form. Typically, the filament used has a diameter of 1.27 mm and this tends to set a limit on how accurately the part can be built.

(2) *Slow process*. The building process is slow, as the whole cross-sectional area needs to be filled with building materials. Building speed is restricted by the extrusion rate or the flow rate of the build material from the extrusion head. As the build material used are plastics and their viscosities are relatively high, the build process cannot be easily speeded up.

(3) *Unpredictable shrinkage*. As the FDM process extrudes the build material from its extrusion head and cools them rapidly on deposition, stresses induced by such rapid cooling invariably are introduced into the model. As such, shrinkages and distortions caused to the model built are a common occurrence and are usually difficult to predict, though with experience, users may be able to compensate for these by adjusting the process parameters of the machine.

4.2.6 Applications

FDM models can be used in the following general applications areas:

(1) *Models for conceptualization and presentation*. Models can be marked, sanded, painted and drilled and thus can be finished to be almost like the actual product.

(2) *Prototypes for design, analysis and functional testing*. The system can produce a fully functional prototype in ABS. The resulting ABS parts have 85% of the strength of the actual molded part. Thus actual testing can be carried out, especially with consumer products.

(3) *Patterns and masters for tooling*. Models can be used as patterns for investment casting, sand casting and molding.

4.2.7   Example

4.2.7.1   *Toyota Uses FDM for Design and Testing [13]*

Toyota, the fourth-largest automobile manufacturer in the United States, produces more than one million vehicles per year. Its design and testing of vehicles are mainly done at the Toyota Technical Center (TTL) USA Inc.

In 1997, TTL purchased the Stratasys FDM 8000 fused deposition modeler (FDM) system to improve on their efficiency in design and testing. The system, not only is able to produce excellent physical properties prototype, it is also able to produce them fast. Furthermore, the system does not require any special environment to be operated in.

In the past, fabricating a prototype was costly and time consuming at TTL. To manufacture a fully functional prototype vehicle, it required $10 000 to $100 000 to manufacture a prototype injection mold and it took as long as 16 weeks to produce. Furthermore, the number of parts required was around 20 to 50 pieces and thus, the conventional tooling method is unnecessarily costly.

In the Avalon 2000 project, TTL replaced its conventional tooling method with the FDM system. Although a modest 35 parts were being replaced by rapid prototypes, it was estimated that it saved Toyota more than $2 million in prototype tooling costs. Moreover, rapid prototyping also helped designers to identify unforeseeable problems early in the design stage. It would have added to the production costs significantly if the problems were discovered during the production stage.

The physical properties of these prototypes are not identical to those made from the conventional method, but nevertheless, as claimed by one of the staff in TTL, they are often good enough. TTL plans to increase its rapid prototyping capacities by introducing additional units of the FDM system. Its aim is to eliminate all conventional prototyping tooling and go straight to production tooling in the near future.

4.2.7.2   *JP Pattern Uses FDM for Production Tooling*

JP Pattern, a prototype and production tooling service bureau in Butler, Wisconsin, was chosen by Ford Motor Company to work with a

Figure 4.7: Investment cast wax patterns for injection mold created using FDM
(Courtesy Stratasys Inc.)

consortium of companies dedicated to finding a faster and less expensive way to produce dunnages [14]. Dunnages are material handling parts used to hold bumpers and fenders in place when they are shipped or used in Ford's production line. By Stratagys' FDM process, the rapid prototypemaster (Figure 4.7) was created in investment casting wax. The wax model was then used to create ceramic shell molds, which were investment cast in A2 steel. By using the FDM process, 50% of the costs on production tooling were saved while achieving a 50% time savings when compared with conventional methods.

4.3 KIRA'S PAPER LAMINATION TECHNOLOGY (PLT)

4.3.1 Company

Kira Corporation Ltd. was established in February 1944 and has developed a wide range of industrial products since then. Its line of products includes CNC machines, automatic drilling/tapping machines, recycle units and systems and folding bicycles. In recent years, Kira has joined the development of rapid prototyping technology and has produced two rapid prototyping machines, the KSC-50N (PLT-A3) and the PLT-A4. These two systems use a technology known as Paper Lamination Technology (PLT), formerly known as Selective Adhesive and Hot Press (SAHP). The speed and reliability of both machines have ranked the systems as one of the best systems. The address of the company is Kira Corporation Ltd., Tomiyoshishinden, Kira-Cho, Hazu-gun, Aichi, Japan.

4.3.2   Products

4.3.2.1   *Models and Specifications*

Kira Corporation produces two models of rapid prototyping systems, Kira Solid Center, KSC-50N and PLT-A4 (see Figure 4.8). The specifications of these two models are summarized in Table 4.4.

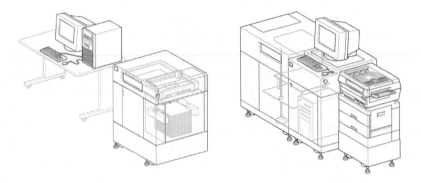

Figure 4.8: Kira's PLT-A4 (left) and KSC-50N (right) (Courtesy of Kira Corporation)

Table 4.4: Specifications of KSC-50N and PLT-A4 (Courtesy of Kira Corporation)

Model	KSC-50N	PLT-A4
Technology	Paper Lamination Technology (PLT)	
Materials	Plain paper and toner	Special paper
Max. work size or build volume, mm	$400 \times 280 \times 300$	$190 \times 280 \times 200$
Accuracy, mm	x, y: ±0.05, z: ±0.10	
Power requirement	200 V, 3 phase, 30 A	100 V, single phase, 15 A + 5 A
System size, mm	$2130 \times 1000 \times 1400$	$850 \times 870 \times 200$
Weight, kg	600	450
Data control unit	IBM compatible PC	

4.3.3 Process

The Kira Solid Center system consists of the following hardware: an IBM-compatible PC, a photocopy printer, a paper alignment mechanism, a hot press, and a mechanical cutter plotter [15]. The Kira Solid Center system makes a plain paper layered solid model. The process is somewhat similar to that of the LOM™ process, except that no laser is used and a flat hot-plate press is used instead of rollers. It is called Paper Lamination Technology (PLT) and formerly known as Selective Adhesive and Hot Press (SAHP), the process includes six steps: generating a model and printing resin powder, hot pressing, cutting the contour, completing the block, removing excess material and post processing (see Figure 4.9).

First, the 3D data (STL files) of the model to be built is loaded onto the PC [Figure 4.9(a)]. The model is then oriented within the system, with the help of the software, for the best orientation for the build [Figure 4.9(b)]. Once this is achieved, the system software will proceed to slice the model [Figure 4.9(c)] and generate the printing data based on the section data of the model [Figure 4.9(d)]. The resin powder or toner is applied on a sheet of paper using a typical laser stream printer and is referred to as the Xerography process (i.e., photocopying). The printed area is the common area of two consecutive sections of the model.

A sheet alignment mechanism then adjusts the printed sheet of paper onto the previous layer on the model (the first layer starts from the table). A hot press then moves over them with the printed sheet pressed to a hot plate at high pressure. The temperature controlled hot press melts the toner (resin powder), which adheres the sheets together [Figure 4.9(e)]. The hot press also flattens the top surface and prevents the formation of air bubbles between the sheets. The PC measures the amount of movement up to the hot plate to compensate for any deviation of the sheet thickness.

The PC then generates plotting data based on the section data of the model. A mechanical cutter cuts the top layer of the block along the contour of the section as well as parting lines from which excess paper are removed later [Figure 4.9(f)]. These steps are repeated until the

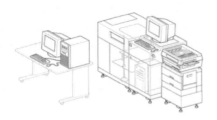

(a) 3D data (STL files) is loaded

(b) Model is oriented within system

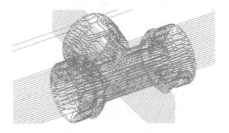

(c) System slices STL data

(d) New paper layer is placed on block

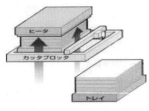

(e) Block and new layer are pressed against hot plate

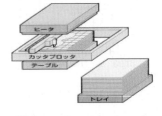

(f) Layer is cut to shape by computerized knife

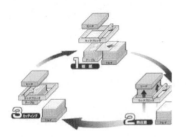

(g) Processes are repeated rapidly

(h) Unnecessary portions are removed

Figure 4.9: Block diagram of PLT process

entire model is built [Figure 4.9(g)]. When printing, hot pressing and cutting are completed, the model block is removed from the machine and unnecessary portions of the paper are disconnected quickly and easily sheet-by-sheet [Figure 4.9(h)]. When the model is complete, its surface may then be finished by normal wood-working or mechanical means.

The tensile strength and bending strength of the material made by the earlier SAHP process has been shown to be approximately one-half that of the wooden models. However, enhancement to the PLT process has enabled model hardness to improve significantly and it has been reported that a hardness of up to 25% better than the equivalent wood model has been achieved.

4.3.4 Principle

The principle of the process is based on the photocopy principle, conventional mechanical layering and cutting techniques. A typical laser stream printer is used for printing and a resin powder instead of print toner is used as toner, which is applied to the paper in the exact position, indicated by the section data to adhere the two adjacent layers of paper.

Three factors, cutter plotter, temperature and humidity, affect the accuracy of the model being built. The accuracy of the cutter plotter affects the accuracy of the model in the X and Y directions. The shrinkage of the model occurs when the model is cooled down in the hot press unit. Expansion of the model occurs when the model is exposed to varying humidity conditions in the hot press unit.

4.3.5 Advantages and Disadvantages

The key advantages in using PLT, as describe by Kira Corporation [16] are as follows:

(1) *Flatness*. The PLT process uses a flat plate and high pressure to bond the layers together. Each layer is pressed with a flat hot plate

and the model remains flat during the entire build process. Since the block is released after cooling, there is minimal internal strain and therefore there is little of no curling in the final model.

(2) *Surface smoothness.* The PLT process uses a computerized knife to cut sheets of paper which result in a smooth surface for the built model. Better surface finishing can be further attained by simple wood working tools but are seldom necessary. Prototypes can be sanded, cut or coated according to the user's needs.

(3) *Hardness.* High lamination pressure used in the PLT process has resulted in products that are 25% harder than equivalent wood and this is often strong enough for most prototyping applications, including sand casting.

(4) *Support structures.* Additional support structures are not necessary in the PLT process as the part is supported by its own material that is outside the perimeter of the cut-path. These are not removed during the hot press process and thus act naturally as supports for delicate and overhang features.

(5) *Office-friendly process.* Kira's machines can be installed in any environment where electricity is available. There is no need for special facilities or utilities for running the KSC or PLT. Moreover, the process is safe as no high-powered laser or hazardous materials are used.

The main disadvantages of using PLT are as follows:

(1) *Inability to vary layer thickness.* Fabrication time is slow in the Z-direction as the process builds prototypes layer by layer, and the height of each layer is fixed by the thickness of the paper used. Thus, the speed of the build cannot be easily increased as the thickness of the paper used cannot be varied.

(2) *Fabrication of thin walls.* Like the LOM™ process, the PLT process is also not well suited for building parts with delicate thin walls, especially walls that are extended in the Z-direction. This is due to the fact that such walls are joined transversely and thus may not be sufficiently strong to withstand the necessary post-processing. However, the flat hot plate used in the process may be

able to pack the heat seal adhesive resin a little better to limit the problem.

(3) *Internal voids.* Models with internal voids cannot be fabricated within a single build as it is impossible to remove the unwanted support materials from within the "void."

(4) *Removal of supports.* Similar to that of the LOM™ process, the most labor-intensive part of the build comes at the end — separating the part from the support material. However, as the support material is not adhesively sealed, the removal process is simpler, though wood working tools are sometimes necessary. The person working to separate the part needs to be cautious and aware of the presence of any delicate parts on the prototype so as not to damage the prototype during post-processing.

4.3.6 Applications

The PLT has been used in many areas, such as the automobile, electric machine and component, camera and office automation machine industries. Its main application area is in conceptual modeling and visualization. In Japan, Toyota Motor Corp., NEC Corp. and Mitsubishi Electric Corp. have reportedly used the Solid Center KSC-50N and its predecessor to model their products.

4.4 3D SYSTEMS' MULTI-JET MODELING SYSTEM (MJM)

4.4.1 Company

3D Systems is the world leader in the rapid prototyping industry where their rapid prototyping machines account for nearly half of the world's machines in use at service companies. The multi-jet modeling system (MJM) was first launched in 1996 as a concept modeler for the office to complement the more sophisticated SLA® machines (see Section 3.1). The company's details are found in Section 3.1.1.

4.4.2   Products

4.4.2.1   *Models and Specifications*

3D systems introduced its 3D printer, the ThermoJet® (see Figure 4.10) in 1999, as a replacement for its old model, Actua™ 2100 that was launched in 1996. Unlike other 3D Systems' rapid prototyping systems, ThermoJet® is intended as a concept modeler. The purpose of a concept modeler is mainly to generate a 3D model in the fastest possible time for design review. A summary of the product specifications is found in Table 4.5.

4.4.3   Process

The process of the ThermoJet® is simple and fully automated. It consists of the following steps.

(1) ThermoJet® uses the ThermoJet® Printer Client Software to input "sliced" STL files from the CAD software. The ThermoJet® Printer Client Software is a powerful software which allows users to verify the preloaded STL files and auto-fix any errors where necessary. The software also helps users to auto-position the parts to be built so as to optimize building space and time. After all details have been finalized, the data is placed in a queue, ready for ThermoJet® to build the model.

(2) During the build process, the head is positioned above the platform. The head begins building the first layer by depositing materials as it moves in the X-direction. As the machine's print head contains a total of 352 heads and measures 200 mm across, it is able to deposit material faster and more efficiently as compared to the older model, Actua™ 2100 (which has only 96 heads).

(3) With a print head measuring 200 mm across, ThermoJet® is able to build a model with a width of up to 200 mm in a single pass. If the model's wide is greater than 200 mm, then the platform is repositioned (Y-axis) to continue building in the X-direction until the entire layer is completed.

Figure 4.10: 3D System's ThermoJet® concept modeler (Courtesy 3D Systems)

Table 4.5: Specifications of ThermoJet® (Courtesy 3D Systems)

Model	ThermoJet® Printer
Technology	Multi-Jet Modeling
Resolution: (x, y, z)	$300 \times 400 \times 600$ DPI (dots per inch)
Max. model size: mm (in)	$250 \times 190 \times 200$ ($10 \times 7.5 \times 8$)
Available build materials	ThermoJet 2000 and ThermoJet 88 thermoplastics build material
Material color options	Neutral, grey, or black
Interface	TCP/IP protocol; Ethernet 10/100 based-TX network; RJ connector required
Platform support	Silicon Graphics IRIX v6.5.2 (open GL required) Hewlett Packard HP-UX v10.2 ACE (open GL) Sun Microsystems Solaris v2.6.0 (Open GL) IBM RS/6000 AIX v4.3.2 (Open GL) Windows NT v4.0, Windows 98, 2000 and ME edition
Power consumption	100 VAC, 50/60 Hz, 12.5 A 115 VAC, 50/60 Hz, 10.0 A 230 VAC, 50/60 Hz, 6.3 A
Dimensions: $w \times d \times h$, m (in)	$1.37 \times 0.76 \times 1.12$ ($54 \times 30 \times 44$)
Shipping dimensions (with accessory kit): m (in)	W1.58 × D1.02 × H1.60 (W62.25 × D40.25 × H63)

(4) After one layer is completed, the platform is lowered and the building of the next layer begins in the same manner as described in Steps 2 and 3.

(5) The process continues with the continual repetition of Steps 2 to 4 until the part is complete, after which the part is ready for instant removal and review with no further need for post-processing or post-curing.

4.4.4 Principles

The principle underlying ThermoJet® is the layering principle, used in most other RP systems. MJM builds models using a technique akin to ink-jet or phase-change printing, applied in three dimensions. A print head comprising 352 jets oriented in a linear array builds models in successive layers, each jet applying a special thermopolymer material only where required. The MJM heads shuttles back and forth like a line printer (X-axis), building a single layer of what will soon be a three-dimensional concept model. If the part is wider than the print head, the platform repositions (Y-axis) itself to continue building the layer. When the layer is complete, the platform is distanced from the head (Z-axis) and the head begins building the next layer. This process is repeated until the entire concept model is complete. The main factors that influence the performance and functionalities of the ThermoJet® are the thermopolymer materials, the MJM head, the $X-Y$ controls, and the Z-controls.

4.4.5 Advantages and Disadvantages

The advantages of the MJM technology used on the ThermoJet® are as follows:

(1) *Efficient and ease of use*. MJM technology is an efficient and economical way to create concept models. The large number of jets from 352 heads allows fast and continuous material deposition at a resolution of 300 dpi for maximum efficiency. MJM builds models directly from any STL file created with any three-dimensional solid

modeling CAD program and no file preparation is required. This makes the concept modeler easy to use.

(2) *Cost-effective*. MJM uses inexpensive thermopolymer material that provides for cost-effective modeling.

(3) *Fast build time*. As a natural consequence of MJM's raster-based design, the geometry of the model being built has little effect on the building time. Model work volume (envelope) is the singular determining factor for part build time.

(4) *Office-friendly process*. As the system is clean, simple and efficient, it does not require special facilities, thereby enabling it to be used directly in an office environment. Due to its networking capabilities, several design workstations can be connected to the machine just like any other computer output peripheral.

The disadvantages of the MJM technology used on the ThermoJet® are as follows:

(1) *Small build volume*. The machine has a comparatively small build volume as compared to most other high-end rapid prototyping systems (e.g., SLA-500), thus only small prototypes can be fabricated.

(2) *Limited materials*. Materials selection are restricted to 3D systems' ThermoJet® 2000 and ThermoJet 88 thermopolymer. This limited range of material means that many functionally-based concepts that are dependant on material characteristics cannot be effectively tested with the prototypes.

(3) *Weak accuracy*. The process lacks sufficient accuracy in building exact prototypes when compared with the high-end RP systems. Nevertheless, these prototypes are often good enough to be used as visualization models or physical representations of the design.

4.4.6 Application

The main application for the ThermoJet® is to produce concept models for visualization and proofing during the early design process. It is meant to function in an office environment in the immediate vicinity of the CAD workstations. Thus it is offered as a complement, producing quick concept models for the SLA® machines, which are meant for producing the precision prototypes and for rapid tooling.

4.4.7   Example

4.4.7.1   *Concept Modeler Speeds Handheld Scanner Development at Symbol Technology [17]*

Symbol Technology of Holtsville, New York, is one of the leading manufacturers for handheld computers and laser bar-code scanners used in various industries like transportation and logistics, in the United States. Due to the stiff competition within the industry, designers and engineers in Symbol Technology need to introduce new products to the market in the shortest possible time in order to reap the benefits of being the first to enter the market.

The company initially installed a high-end rapid prototyping machine, the SLA-250 stereolithography system (also by 3D Systems) to produce all prototypes and engineering models. Although the system yields high accuracy and precision, the turnaround time to build the prototype is still substantial. Furthermore the model needs considerable post processing before it can be used.

The engineering staff and designers found that the time required was too long especially when the company wants to introduce a new product fast. Thus the company purchased another machine, the ThermoJet® modeler from 3D systems to remedy the problem.

The ThermoJet® modeler is able to speed up the whole building time of these concept models, from about a week for stereolithography models to just a few hours. One advantage of ThermoJet® over SLA-250 is the former does not require any post curing process and the supports created during the process can be easily removed. Furthermore, the system does not require many people to operate it. Symbol Technology employs only one person to operate both the ThermoJet® modeler and the SLA-250 system.

The introduction of the ThermoJet® concept modeler into the company has fundamentally changed the whole design process in the company as prototypes can be produced quickly.

However, not all of Symbol's engineers prefer to use the ThermoJet® system. One main reason for choosing the high-end rapid prototyping system is that the system is capable to produce models with greater durability. Thus, the choice of which system to use is greatly influenced by the task the prototype is supposed to perform. If prototypes are used merely as a visual aid, then the ThermoJet® modeler becomes the natural choice.

4.5 SOLIDSCAPE'S MODELMAKER AND PATTERNMASTER

4.5.1 Company

Solidscape, formerly known as Sanders Prototype Inc., was established in 1994. Solidscape is a privately held company and was funded by Sanders Design Inc. (SDI), a research company, which is principally owned and managed by Solidscape's founder, Royden C. Sanders, to research and market the ModelMaker™ rapid prototyping system. Solidscape's machines are based on ink jet technology, similar to the technology used by Sander Design Inc., and are installed in over 14 countries around the world. The address of the company is Solidscape Inc., P. O. Box 540, Pine Valley Mill, Wilton, New Hamphire 03086, USA.

4.5.2 Products

4.5.2.1 *Models and Specifications*

Solidscape currently produces two rapid prototyping machines in the market, namely the ModelMaker II™ and the PatternMaster™ (see Figure 4.11). These two machines are classified as concept modeler but from the point of view of their capabilities, they are capable of functioning as more than a concept modeler. These machines are also capable of producing tooling patterns that are suitable for casting or mold making.

Figure 4.11: Solidscape's PatternMaster™ (Courtesy Solidscape Inc.)

The ModelMaker™ and the PatternMaster™ are high precision machines. They can achieve, an accuracy of 0.025 mm per mm (or 0.001 inch per inch) in all X, Y and Z directions and an excellent surface finish of 0.8–1.6 μm (32–63 micro inches) RMS. Both machines are able to build prototypes up to a maximum volume of 304.8 × 152.4 × 213.9 mm (12 × 6 × 8.5 inches). However both systems are limited to using only ProtoBuild and ProtoSupport, Solidscape's proprietary thermoplastics materials.

The supports generated can be easily removed by immersing the built model in the BIOACT™ VSO bath at 54°C–76°C (122°F–158°F). Although both systems have almost the same specifications, Solidscape claims that the PatternMaster™ is nearly twice as fast as the ModelMaker™ [18]. The specifications for both systems are summarized in Table 4.6.

4.5.3   Process

The process of the ModelMaker II™ and PatternMaster™ can be considered as a hybrid of FDM and 3D Printing. Solidscape Inc. calls it "3D Plotting" [19]. The process uses two ink-jet type print heads, one depositing the thermoplastic building material and the other depositing supporting wax. Like most other commercial systems, the model is built

Table 4.6: Specifications of ModelMaker™ and PatternMaster™
(Courtesy Solidscape Inc.)

Model	ModelMaker II	PatternMaster
Process	Ink-jet-based technology	
Build envelop, mm (in)	304.8 × 152.4 × 215.9 (12 × 6 × 8.5)	
Layer thickness, mm (in)	0.013–0.076 (0.0005–0.0030)	
Accuracy, mm (in)	±0.025 (±0.001)	
Surface finish (RMS)	0.8–1.6 μm (32–63 micro-inches)	
Min. feature size, mm (in)	0.254 (0.010)	

on a platform, which is lowered one layer after each layer is built. The liquefied build material cools as it is ejected from the print head and solidifies upon impact on the model. Wax is deposited to provide a flat, stable surface for deposition of build material in the subsequent layers. After each layer is completed, a cutter removes approximately 0.025 mm off the layer's top surface to provide a smooth, even surface for the next layer.

After several layers have been deposited, print heads are moved automatically to a monitoring area to check for clogging of the jet nozzles. If there is no blockage, the process continues, otherwise the jets are purged and the cutter removes any layers deposited since the last check of jets and restarts the building process from that point.

4.5.4 Principle

The principle underpinning Solidscape's ModelMaker II™ and PatternMaster™ is the layering principle, like in most other RP systems. The systems build models using a technique akin to an ink-jet, applied in three dimensions. Two print heads, one applying a special thermopolymer material and the other depositing supporting wax at positions where required. The twin heads shuttles back and forth much like a line printer but only in three-dimension to build one layer. When the layer is complete, the platform is distanced from the head (Z-axis) and the head begins building the next layer. This process is repeated until the entire concept model is completed. The main factors that influence the performance and functionalities of the ModelMaker II™ and PatternMaster™ are the thermopolymer materials, the ink-jet print head, the X–Y controls, and the Z-controls.

4.5.5   Advantages and Disadvantages

The main advantages of using ModelMaker II™ and PatternMaster™ are as follows:

(1) *High precision.* The systems are able to achieve an accuracy of ±0.025 mm per mm (±0.001 inch per inch) in the X, Y and Z directions. They are also able to produce a considerably smooth surface finish, 0.8–1.6 μm (32–63 micro-inches) RMS. This compares favorably to many other concept modelers (and even RP machines).

(2) *Adjustable build layers.* Build layers can be adjusted, ranging from 0.013 mm (0.0005 inch) to 0.076 mm (0.0030 inch) depending on the users' requirements and the need to speed up the build process.

(3) *Office-friendly process.* These systems can be used in an office environment, without any special facilities as the process is simple, clean and efficient. Moreover, materials used are nontoxic, which makes the process safe for office operation.

(4) *Minimal post-processing.* Prototypes fabricated using these systems require minimal post-processing and they are sufficiently strong and can be used directly for casting.

The disadvantages of using ModelMaker II™ and PatternMaster™ are as follows:

(1) *Small build volume.* The build volume of both systems are comparatively smaller than many high-end rapid prototyping systems (only 304.8 × 152.4 × 213.9 mm). Thus the systems are not suitable for building prototypes that are larger than these dimensions.

(2) *Limited materials.* Like many other concept modelers, materials that can be processed by these systems are restricted to

Solidscape's ProtoBuild™ and ProtoSupport™. Thus it puts a limitation to the range of material-dependant functional capabilities of the prototypes.

(3) *Slow process*. Although PatternMaster™ is twice as fast as ModelMaker II™, both systems are still considerably slower when compared to other major rapid prototyping systems.

4.5.6   Application

The main application for Solidscape's systems is to produce concept models for visualization and proofing during the early design process. The machines are intended to function in an office environment in the immediate vicinity of the CAD workstations. These machines are also able to build tooling patterns intended for rapid tooling.

4.5.7   Example

4.5.7.1   *Grotell Design Uses ModelMaker II™ in Design Process [20]*

Grotell Design Inc., New York, USA, is one of the most successful design companies in the United States. The company's direction is committed to the design and development of fine watches and timepieces product that embody the unique identity and culture of its clients. Due to the high quality and fine design of their products, there is a need for a high precision system to match the company's requirement. The system needs also to help designers to study forms and also to verify engineering assumptions for the development of their signature products. Several systems were tested and they proved not to be favorable with regards to cost and resolution.

After several testing, Grotell Design chose Solidscape's ModelMaker II™ system. The system, as claimed by the company, is the only system that incorporates the technology that provides the ability to generate a highly precise pattern for every major component of the watch design. Furthermore, the system is easy to use and operate. Young designers from the company need only a good 45 minutes to overview on the system before they are comfortable in setting up and running the system.

4.6   BEIJING YINHUA'S SLICING SOLID MANUFACTURING (SSM), MELTED EXTRUSION MODELING (MEM) AND MULTI-FUNCTIONAL RPM SYSTEMS (M-RPM)

4.6.1   Company

Beijing Yinhua Rapid Prototypes Making and Mold Technology Co. Ltd. and the Center for Laser Rapid Forming (CLPF), Tsinghua University is one of the few joint ventures between universities and companies in the People's Republic of China to research on and develop rapid prototyping machines. This joint venture, under the guidance of Professor Yan Yongnian of Tsinghua University and support by the National Natural Science Foundation of China (NNSFC), has seen the company launched a series of rapid prototyping systems in just a few short years. The earliest model, which is believed to be the first rapid prototyping machine made in China, was the Multi-functional Test System (M-220). The machine was built in early 1994 and is currently still in use in CLPF, Tsinghua University. The address to the company is Beijing Yinhua Rapid Prototypes Making and Mold Technology Co. Ltd., Department of Mechanical Engineering, Tsinghua University, Beijing 100084, People's Republic of China.

4.6.2   Products

4.6.2.1   *Models and Specifications*

Beijing Yinhua has developed three different kinds of rapid prototyping systems. These are the Slicing Solid Manufacturing (SSM), the Melted Extrusion Modeling (MEM) and the Multi-functional Rapid Prototyping Manufacturing (M-RPM) machines. A summary of the product specifications is found in Table 4.7.

4.6.2.2   *Slicing Solid Manufacturing*

The SSM uses a technique very similar to Cubic Technology's LOM™ (Laminated Object Manufacturing®) where a CO_2 laser is used to cut

Table 4.7: Specifications of Beijing Yinhua's rapid prototyping systems
(Courtesy Center for Laser Rapid Forming, Tsinghua University)

Model	SSM-600	SSM-1600	MEM-250-II	M-RPMS-II	
Technique	Slicing Solid Manufacturing		Melted Extrusion Manufacturing	Slicing Solid Manufacturing	Melted Extrusion Manufacturing
Maximum build volume, mm	$600 \times 400 \times 500$	$1600 \times 800 \times 700$	$250 \times 250 \times 250$	$600 \times 400 \times 500$	$500 \times 400 \times 500$
Precision, mm	~0.10	~0.15	~0.15	~0.10	~0.15
Scanning speed, mm/sec	0~500	0~500	0~500	0~500	0~400
Laser system	40 W, CO_2	40 W, CO_2	N.A.	40 W, CO_2	N.A.
Temperature variation	N.A.	N.A.	±1°C	N.A.	±1°C
Materials	Sheets with binders	Sheets with binders	Waxy, ABS and nylon filaments	Sheets with binders	Waxy, ABS and nylon filaments

the shape of each layer (see Section 4.1). However, according to the company, SSM has its own special features [21]. The company currently offers two versions of SSM, the SSM-600 and the SSM-1600. Beijing Yinjua claimed that the SSM-1600 has the largest built volume in the world, measuring $1600 \times 800 \times 750$ mm (63" × 32" × 30").

4.6.2.3 *Melted Extrusion Modeling*

The second system is the MEM (see Figure 4.12) and it is very similar to Stratasys's FDM (Fused Deposition Manufacturing) process (see Section 4.2). However, the system has its own distinguishing features [21]. The MEM process heats and extrudes thermoplastics in filament

Figure 4.12: Beijing Yinhua's MEM rapid prototyping machine
(Courtesy Center for Laser Rapid Forming, Tsinghua University)

form onto a fixed base through a three axis extrusion head. The head
has the capability of moving upwards between layers. The available
model is MEM-250-II and it has a forming space of 250 × 250 ×
250 mm (10 × 10 × 10 in).

4.6.2.4 *Multi-functional RPM System*

A third system developed by Beijing Yinhua is called the Multi-
functional Rapid Prototyping Manufacturing (M-RPM). The distinct
feature of this system is that it combines both features of the SSM and
MEM processes into a single machine. Currently, the company offers
one model of the rapid prototyping system, the M-RPMS-II. The
M-RPMS-II is considered more efficient and reliable as compared to
the earlier nonproduction version of the M-RPMS.

4.6.3 Processes

4.6.3.1 *Slicing Solid Manufacturing*

The SSM process is very similar to the LOM™ process discussed in
Section 4.1.3. The CAD file of the part is imported into the system and

processed by the resident software called Lark '97 which slices the part for the build. Parts are built, layer-by-layer, by laminating sheet-form materials (mainly paper) and the contour of the part is cut by a powerful CO_2 laser. The heat sealing of the binder (adhesive) is by means of a rapidly heated flat hot plate instead of rollers. The elevation platform, which holds the resultant part and "block" of support lowers when each layer is completed to be ready for the next layer. The exact Z-height is measured so that the correct cross-section slice data can be computed for the next layer. Additional support structures are not necessary as the excess sheet material forms the natural support of the part.

4.6.3.2 *Melted Extrusion Modeling*

The MEM process is similar to that of Stratasys' FDM as described in Section 4.2.3. The CAD file of the part is imported into the system and processed by the same resident software, Lark '97, which is also used in the SSM. Thermoplastics or wax in filament form is melted and extruded to the model via a three-axis (X, Y, Z)-controlled extrusion head. The model is built on a fixed platform. While the model is built layer-by-layer, the three-axis controlled extrusion head is able to move between layers, giving users added capabilities.

4.6.3.3 *Multi-functional RPM System*

The process of the M-RPM system is based on the combination of the SSM and MEM systems into a single machine to take advantage of these two processes. The notion is that the processing software (Lark '97) for slicing, filling and supporting the model and the control software of both SSM and MEM processes are common. Also the machine structure and platform that holds the model can also be combined, thus this combination machine will be able to produce either SSM or MEM models, depending on the need of the user. The process of building either model is the same as described in Sections 4.6.3.1 and 4.6.3.2.

4.6.4   Principle

The principle used for Beijing Yinhua's machines are very similar to those describe for LOM™ and FDM for SSM and MEM respectively (see Sections 4.1.4 and 4.2.4). Like all other RP machines, it utilizes the layer-by-layer method of building the model, with a small difference between the systems in that for the SSM, the platform lowers while for MEM, the platform remain fixed. One notable variance of the M-RPM system as compared to most other RP systems is that by combining the two systems (SSM and MEM) within a single platform, customers are able to have a machine with dual capabilities in building both laminated models and deposition models. The company hopes that this will provide more options to their customers wishing to enhance their own RP capabilities in providing better prototyping services.

4.6.5   Advantages and Disadvantages

The advantages and disadvantages of the systems are summarized in the Tables 4.8 and 4.9 respectively.

4.6.6   Applications

The SSM, MEM and M-RPM have been used in many Chinese industries, including automobile, electric machine, and consumer electronics industries. The main application is in conceptual modeling and visualization. Modified version of the MEM have been used in biomedical applications using poly-lactic acid (PLA) to build a segment of a bone for a rabbit leg in experiments.

4.7   CAM-LEM's CL 100

4.7.1   Company

CAM-LEM (Computer Aided Manufacturing of Laminated Engineering Materials) Inc., a privately owned company in the US, commercializes a rapid prototyping technology akin to Cubic Technology's Laminated

Table 4.8: Advantages of Beijing Yinhua's rapid prototyping systems

System	Advantages
SSM	(1) *Fast build time.* The time needed to cut the layer to the desired shape is fast, as the laser does not have to scan the entire area of the cross-section but just cut around its periphery. (2) *Support structure.* No support structure is required as the part is supported by its own material. (3) *Post-curing.* Parts can be used immediately after the process and no post-curing is required.
MEM	(1) *Minimal wastage.* System does not waste materials during or after producing the model and it does not require clean up. (2) *Ease of material change.* Material is supplied in spool form which is easy to handle and can be easily changed.
M-RPM	(1) *Wider process selection.* The system incorporates two processes (SSM and MEM). Users can choose either of these two processes.

Table 4.9: Disadvantages of Beijing Yinhua's rapid prototyping systems

System	Disadvantages
SSM	(1) *Precise power adjustment.* Power needs to be precisely adjusted so that the laser do not penetrate into previous layers and distort the whole prototype. (2) *Integrity of prototypes.* Integrity of the built prototypes is limited to the bonding strength of the glued layer. (3) *Removal of supports.* The process requires users to manually remove the supports, which can be tedious and time consuming.
MEM	(1) *Mediocre accuracy.* Accuracy is restricted due to the shape of the material used and is limited to the diameter of filament used. (2) *Slow process.* The building process is slow, as the whole cross-sectional area needs to be filled with building materials. (3) *Unpredictable shrinkage.* Rapid cooling in the process causes shrinkage and distortion of models which are hard to predict.
M-RPM	(1) As M-RPM uses both SSM and MEM processes, the disadvantages are similar as the above.

Object Manufacturing® (LOM™). The process uses the "form-then-bond" laminating principle where the contour of the cross section is first cut before laminating it to the previous layers. This technology is developed by a team of researchers from Case Western Reserve University. The company's strategy is not to sell the developed rapid prototyping machine but rather, CAM-LEM intends to act as a service provider to interested customers. The company's address is CAM-LEM Inc., 540 East 105th street, Cleveland, Ohio 44108, USA.

4.7.2   Products

4.7.2.1   *Models and Specifications*

The rapid prototyping systems developed by CAM-LEM Inc. is called CL-100 and it is able to produce parts up to 150 × 150 × 150 mm in size. One distinct feature of CL-100 is that the system is able to build metal parts and ceramic parts with excellent mechanical properties after the sintering process. The layer thickness ranges from 100 to 600 microns and higher, and up to five different materials can be incorporated into a single automated build cycle.

4.7.3   Process

The CAM-LEM approach (see Figure 4.13), like other rapid prototyping methods, originates from a CAD model decomposed into the boundary contours of thin slices. In the CAM-LEM process these individual slices are laser cut from a sheet stock of engineering material (such as "green" ceramic tape) as per the computed contours. The resulting part-slice regions are extracted from the sheet stock and stacked to assemble a physical 3D realization of the original CAD description. The assembly operation includes a tacking procedure that fixes the position of each sheet relative to the pre-existing stack. After assembly, the layers are laminated by warm isostatic pressing (or other suitable method) to achieve intimate interlayer contact, promoting high-integrity bonding in the subsequent sintering operation. The laminated "green" object is then fired (with an optimized heating schedule to

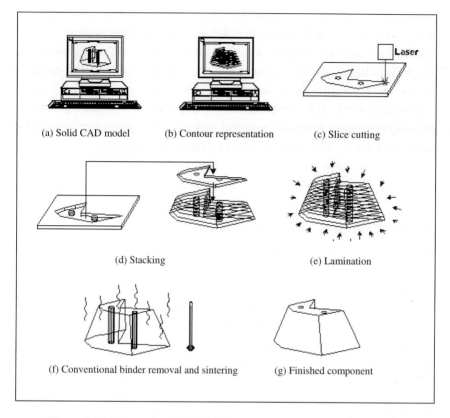

(a) Solid CAD model   (b) Contour representation   (c) Slice cutting

(d) Stacking   (e) Lamination

(f) Conventional binder removal and sintering   (g) Finished component

Figure 4.13: Schematic of CAM-LEM process (Courtesy CAM-LEM Inc.)

densify the object and fuse the layers and particles within the layers) into a monolithic structure. The result is a 3D part, which exhibits not only the correct geometric form, but functional structural behavior as well.

4.7.4 Principle

The CAM-LEM process is based on the same principle applied to most solid-based RP systems. Objects are built layer-by-layer and the laminated objects can be fabricated from a wide variety of engineering materials. The sections that form the layers are cut separately from sheet stock with a CO_2 laser and are then selectively extracted and precisely stacked. Multiple material types can be used within a single build. This process allows for the formation of interior voids and channels without manual waste removal, thus overcoming the problem of entraped volume that plague most other RP systems. The distinct characteristic of this process is the separation of the geometric formation process from the material process, thus providing users with more flexibility. The crucial factors affecting the quality of the built model are the laser cutting, the indexing and tacking, the alignment of the stacking process, the binding process and the sintering process.

4.7.5   Advantages and Disadvantages

The key advantages of using CAM-LEM technology are as follows:

(1) *Elimination of interior voids and channels.* The CAM-LEM process separates the laser cutting process from the stacking and lamination process, thus allowing for the formation of interior voids and channels, thereby eliminating the problem of entraped volumes that troubled many other RP systems.

(2) *Laser power adjustment.* The cutting laser power of CAM-LEM technology does not need to be precisely adjusted because the process uses the "form-then-bond" laminating principle, where the contour of the cross section is first cut before laminating to the previous layers. Thus, it eliminates the problem of the laser burning into the previous layers.

(3) *High-quality prototypes.* As the technology uses the "form-then-bond" principle, it ensures that the layers are free from fine grains of unwanted materials before bonding them to the previous layer. This is highly desirable as the unwanted materials trapped in between layers would affect the mechanical property of the final product and such a situation should be eliminated so as to fabricate high quality prototypes.

(4) *Adjustable build layers.* The CAM-LEM process allows prototypes to be built using different material thicknesses, which could effectively speed up the process. Regions with large volume are

built with thicker sheets of paper and surfaces that require smooth surface finishes are built with thinner sheets of paper.

The disadvantages of using CAM-LEM technology are as follows:

(1) *Significant shrinkage.* The main disadvantage in using CAM-LEM's technology is that prototype will shrink in size by around 12–18%, which makes the dimensional and geometric control of the final prototype difficult.

(2) *Precise alignment.* The process requires high accuracy from the system to align the new bonding layer to the previous layer before bonding it. Any slight deviation in alignment from previous layers will not only affect the accuracy of the model, but also its overall shape.

(3) *Lacks natural supports.* While the process eliminates the problem of entraped volume, it does require users to identify the locations of supports for the prototype, especially for overhanging features. As this process transfers only the desired layers to bond with the previous layers, all unwanted materials, which could be used as natural supports are thus left behind.

4.7.6   Applications

The CAM-LEM process has been used mainly to create rapid tools for manufacturing. Functional prototypes and even production of ceramic and metal components have been built.

4.8   ENNEX CORPORATION'S OFFSET FABBERS

4.8.1   Company

Ennex Corporation was started as Ennex Fabrication Technologies in 1991 by a successful physicist and computer entrepreneur, Mr. Marshall Burns. Then, Mr. Burns was joined by a team of experience and enthusiastic professionals to develop a range of state-of-the-art rapid prototyping machines, which focused mainly on digital manufacturing technology. The company named these machines "digital fabricator" or simply "fabber." A "fabber", which the company defines, is a *"factory in a box" that makes things automatically from digital data* [22]. Besides developing this new technology for digital manufacturing, the company also provides valuable consulting to related companies that deal with "fabbers." The address to the company is Ennex Corporation, 11465 Washington Place, Los Angeles, California 90066, USA.

4.8.2   Products

4.8.2.1   *Models*

One of the "fabbers" which the company has developed is the Genie® Studio Fabber, based on a technology known as "Offset™ Fabbing". The technology is somehow similar to that of KIRA Corporation's Paper Laminating Technology (PLT). The main difference between the two technologies is that "Offset™ Fabbing" uses the "form-then-bond" fabrication [23] method. The working principle of the "form-then-bond" fabrication method is simple. It uses a mechanical knife to cut the outline of the layer and after the sheet of material is cut to the required shape, it is then transferred and laminated onto the previous layer. The process of cutting and laminating of the materials is repeated until the whole model is built layer by layer. Ennex Corp. first announced the "Offset™ Fabbing" technology in 1996's, the company is still working to commercialize its first Genie® Studio Fabber. However, a working prototype has been built in the company's development laboratory and work is underway to fine-tune the "fabber" so as to achieve a high quality of output from the system. Ennex claimed that its output will be fast, up to ten or more times faster than leading systems.

4.8.3   Process

The process of Offset™ Fabbing can be described as follows (see Figure 4.14):

(1) Thin fabrication material (any thin film that can be cut and bonded to itself by means of adhesive) is rested on a carrier.

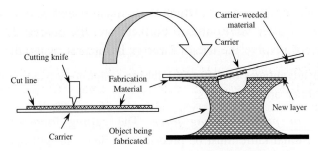

Figure 4.14: Schematic of Ennex Offset™ Fabbing process [23]

(2) A two-dimensional plotting knife is used to cut the outlines of cross sections of the desired object into the fabrication material without cutting through the carrier.

(3) The plotter can also cut parting lines and outlines for support structures.

(4) The film is then "weeded" to remove some or all of the "negative" material.

(5) The film is inverted so that the carrier is facing up and the cut pattern is brought into contact with the top of the growing object and bonded to it.

(6) The carrier is then peeled off to reveal the new layer just added and a fresh surface ready to bond to the next layer.

4.8.4   Advantages and Disadvantages

The advantages of Offset™ Fabbing technology used on the Genie Studio Fabber are as follows:

(1) *Wide variety of materials.* The process is not restricted to a few materials. Basically, any thin film that can be cut and bonded to itself can be used for the process. Paper and thermoplastic sheets are typical examples.

(2) *Minimal shrinkage.* As the bonding layers do not experience any significant change in temperature during fabricating process, the shrinkage of the built parts is minimal.

(3) *Office-friendly process.* The process is clean and easy to use in an ordinary office environment. Furthermore, the process is safe as it does not involve the use of laser or hazardous materials.

The disadvantages of Offset™ Fabbing technology used on the Genie Studio Fabber are as follows:

(1) *Need for precise cutting force.* The cutting force of the two-dimensional knife must be precise, as a deep cut will penetrate into and cut through the carrier while a shallow cut will not give a clean cut to the layer resulting in tearing and sticking.

(2) *Need for precise alignment.* The process requires high accuracy from the system to align the new bonding layer to the previous layer before bonding it. Precise position is critical as any slight deviation in alignment from previous layers will not only affect the accuracy of the model, but also the overall shape of the model.

(3) *Wastage of materials.* The weeded (unwanted) materials cannot be re-used after the layers have been bonded with the previous layer and hence a significant amount of materials are wasted during the process.

4.8.5   Applications

The Offset™ Fabbing process can be used to mainly create concept models for visualization and proofing during the early design process. The fabbers are intended to function in an office environment in the immediate vicinity of the CAD workstations. Functional prototypes and tooling patterns intended for rapid tooling can also be built.

4.9   THE SHAPE DEPOSITION MANUFACTURING PROCESS

4.9.1   Introduction

While most rapid prototyping processes based on the discretized layer-by-layer process are able to build almost any complex shape and form, they suffer from the very process of discretization in terms of surface finish as well as geometric accuracies. The Shaped Deposition Manufacturing process (SDM), first pioneered by Prof. Fritz Prinz and his group at Carnegie Mellon University and later Stanford University, is a rapid prototyping process that overcomes these difficulties by combining the flexibility of the additive layer manufacturing process with the precision and accuracy attained with the subtractive CNC machining process. Shot peening, microcasting, and a weld-based material deposition process can be further combined within a CAD-CAM environment using robotic automation to enhance the capability of the process [24]. The SDM process, though well researched and having many capabilities, is yet to be commercialized at the time of print.

4.9.2 Process

The SDM process is a rapid prototyping process that systematically combines the advantages of layer-by-layer manufacturing with the advantages of precision material removal process [25]. The process can be illustrated by Figure 4.15.

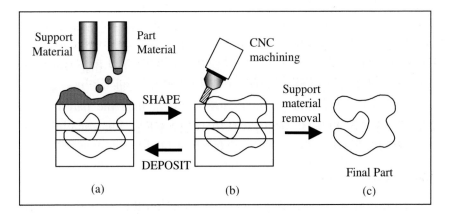

Figure 4.15: Shape Deposition Manufacturing building process

Materials for the individual segments of the part are first deposited at the deposition station to form the layer of the part [See Figure 4.15(a)]. One of the several deposition processes is a weld-based deposition process called microcasting [24], and the result is a near-net shape deposition of the part for that layer. The part is then transferred to the shaping station, usually a five-axis CNC machining center where material is removed to form the desired or net shape of the part [Figure 4.15(b)]. After the shaping station, the part is transferred to a stress relief station, such as shot-peening, to control and relieve residual stress buildup due to the thermal process during deposition and machining. The part is then transferred back to the deposition station where complementary shaped, sacrificial support material is deposited to support the part. The sequence of depositing part or support material is dependent on the geometry of the part as explained in the following paragraph. The process is repeated until the part is complete, after which, the sacrificial support material is removed and the final part is revealed [Figure 4.15(c)].

One major difference that SDM has, unlike most rapid prototyping processes, is that the CAD model of the part is decomposed into slices or segments that maintain the full three-dimensional geometry of the outer surface. The major advantage of this strategy of decomposing shapes into segments, or "compacts" as called by the inventors, is eliminating the need to machine undercut features. Instead, such features are formed by depositing support or part material as appropriate onto previously deposited and shaped segments [see Figure 4.15(a)]. For example, if the part (like in a "V"-shape) overhangs the support, then the support material for that layer will be deposited first. Conversely, if the part is clear of the support in the upward direction [as in Figure 4.15(b)], then the part material for the layer will be deposited first. As such, this flexibility of sequencing part-support material deposition totally eliminates the need to machine difficult undercut features. The total layer thickness is not set arbitrarily but is dependent on the local geometry of the part and the deposition process constraints. Because of the alternating deposition-shaping sequences, disretization steps which plague many other rapid prototyping systems are also eliminated.

Microcasting is a nontransferred MIG (Metal Inert Gas) welding process, which deposits discrete, super-heated molten metal droplets to form dense metallurgical-bonded structures. Apart from the micro-casting deposition process, several other alternative processes are also available to deposit a variety of materials in SDM. These processes and their processing materials are summarized in Table 4.10 [26].

4.9.3 Advantages and Disadvantages

The advantages of Shape Deposition Manufacturing are as follows:

(1) *Wide variety of materials.* The process is capable of handling a wide variety of materials, including stainless steel, steel alloys, metals, thermoplastics, photocurable plastics, waxes, ceramics, etc. There is limitation or constraint on the type of materials that this process can handle.

(2) *Ability to build heterogeneous structures.* In addition to the rapid prototyping of complex shapes, the SDM process is also able to fabricate multi-material structures and it also permits pre-fabricated components to be embedded within the built shapes. These provide the process with the unique capability of fabricating heterogeneous structures which will enable the manufacture of novel product designs.

(3) *Variable layer thickness.* As the CAD model of the part is decomposed into slices or segments that maintain the full three-dimensional geometry of the outer surface, the layer thickness varies. The actual layer thickness will depend on the local geometry of the part and the deposition process constraints.

(4) *Direct creation of functional metal shapes.* The process is one of the only few that are able to directly create or fabricate functional metal shapes with secondary processes needed in other RP systems.

(5) *Ease of creating undercut features.* The decomposition strategy used in the SDM process overcomes the many difficulties of building undercuts that constrain many other RP systems. As such, highly complex shapes and forms can be built with little problem.

Table 4.10: Deposition process on the SDM [26]

Deposition process	Description	Part materials	Support materials
Microcasting	An arc is established between a tungsten electrode of a MIG system and a feedstock wire that is fed from a charged contact tip. The wire melts in the arc, forming a molten droplet which, on accumulating sufficient molten material, falls from the wire to the part. The droplets remain superheated and have enough energy to locally remelt the part to form metallurgical bonding on solidification. A laminar curtain of shielding gas prevents oxidation.	Stainless steel	Copper
Extrusion	Materials are deposited with an extruder with a single screw drive.	Thermoplastics, ceramics	Water-soluble thermoplastics
Two-part resin system	Polyurethanes and epoxy resins are deposited as two-part resin-activator systems.	Polyurethanes, epoxy resins	Wax
Hot wax dispensation	Waxes are deposited with a hot-melt extrusion system. Waxes can be used as either part or support material	Wax	Wax, water-soluble photocurable resins
Photocurable resin desperation	Photocurable resins are deposited with a simple syringe pumping system. This is usually used as support material for wax parts.		Water-soluble photocurable resins
MIG welding	MIG welding is used to deposit directly onto the part for fast deposition.	Steel alloys	Copper
Thermal spraying	Thermal spraying is used to spray thin layers of "high-performance" materials. Plasma sprayers are typically used.	Metals, plastics, and ceramics	

The disadvantages of Shape Deposition Manufacturing are as follows:

(1) *Need for precise control of automated robotic system.* The process requires moving the part from station to station (deposition, shaping, shot-peening, etc.). Such a process demands precise control in placing the part accurately for each built layer, whether deposition, shaping, or shot peening. Invariably, because of the movement, errors can accumulate over several built thicknesses and thus affect the overall precision and accuracy of the part. Also the control of the deposition process has to be precise to prevent creation of voids and causing excess remelting.

(2) *Thermal stresses due to temperature gradient.* The microcasting process of melting metals and precisely depositing them on the part to solidify with the previous layer can result in a steep temperature gradient that introduces thermal stresses into the part and can result in distortion. Controlled shot peening can alleviate the problem but cannot entirely eliminate it. A careful balance of the temperature gradient, the associated internal stress built up, the thermal energy accumulation and their relief has to be achieve in order for SDM to produce quality precision parts.

(3) *Need for a controlled environment.* As several deposition and shaping processes can be incorporated into the SDM, the environment within which the system function has to be controlled, e.g., MIG welding and other hot work can result in waste gases discharge which has to be dealt with.

(4) *A large area is required for the system.* Due to the multiple capability of the SDM process and the several stations (deposition, shaping, etc) needed to accomplish the building of the parts, a relatively large area is necessary for the entire system. This would mean that the system is not suitable for any office environment.

4.9.4 Applications

With its capability to handle multiple materials and deposition processes, the SDM can be applied to many areas in many industries, especially in the building of finished parts and heterogeneous products.

One example of finished parts is the direct building of ceramic silicone nitrite component for aircraft engines which was tested and survived on a jet engine rig at up to 1250°C. The SDM process has also successfully fabricated heterogeneous products like an electronic device by building a nonconductive housing package and simultaneously embedding and interconnecting electronic components within the housing. This is especially useful in fabricating purpose-build devices for special applications like the wearable computer. Many interesting heterogeneous products with multi-materials embedded within the product have also been tested [26].

REFERENCES

[1]   Wohlers, T., *Wohlers Report 2000: Rapid Prototyping and Tooling State of the Industry*, Wohlers Association Inc., 2000.

[2]   Feygin, M., "Apparatus and method for forming an integral object from laminations," U.S. Patent No. 4,752,352, 6/21/1988.

[3]   Feygin, M., "Apparatus and method for forming an integral object from laminations," European Patent No. 0,272,305, 3/2/1994.

[4]   Feygin, M., "Apparatus and method for forming an integral object from laminations," U.S. Patent No. 5,354,414, 10/11/1994.

[5]   Burns, M., *Automated Fabrication*, Prentice Hall, 1993.

[6]   Pak, S.S. and Nisnevich, "Interlaminate strength and processing efficiency improvements in laminated object manufacturing," *Proceedings 5th International Conference Rapid Prototyping*, Dayton, OH, 1994, pp. 171–180.

[7]   Sakid, B., "Rapid Prototype Casting (RPC): The fundamentals of functional metal parts from rapid prototyping model using investment casting," *Proceedings 5th International Conference Rapid Prototyping*, Dayton, OH, 1994, pp. 294–300.

[8]   Industrial Laser Solutions, *Application Report, NASA Builds Large Parts Layer by Layer* [Online]. Available: http://www.industrial-lasers.com/archive/1999/05/ 0599fea3.html, 2001.

[9] Wohlers Report 2001, *Industrial Growth, Rapid Prototyping and Tooling State of the Industry*, Wohlers Association Inc.

[10] Rapid Prototyping Report, *Stratasys's New Insight Software*, 11(3), CAD/CAM Publishing Inc., March 2001: 7.

[11] Rapid Prototyping Report, *Stratasys Introduces Genisys Concept Modeller*, 6(3), CAD/CAM Publishing Inc., March, 1996: 1–4.

[12] Crump, S., "The extrusion of fused deposition modeling," *Proceedings 3rd International Conference Rapid Prototyping*, 1992, pp. 91–100.

[13] Rapid Prototyping Report, *Toyota Uses Fused Deposition Modeling to Bypass Prototype Tooling*, 9(8), CAD/CAM Publishing Inc., September, 1999: 1–3.

[14] Strata Brief, *Investment Casting, FDM User: JP Pattern*, Application: Production Tooling, 1995.

[15] Inui, E., Morita, S., Sugiyama, K., and Kawaguchi, N., *SHAP — A Plain 3D Printer/Plotter Process (1994)*: 17–26.

[16] Kira Solid Center, *Characteristics: Why PLT is So Good* [Online]. Available: http://www.kiracorp.co.jp/EG/pro/rp/RP_characteristic. html, 2001.

[17] Rapid Prototyping Report, *Concept Modeler Speeds Handheld Scanner Development*, 11(2), CAD/CAM Publishing Inc., February 2001: 1–2.

[18] Rapid Prototyping Report, *Sander Rapid PatternMaster*, 10(5), CAD/CAM Publishing Inc., May 2000: 6.

[19] Rapid Prototyping Report, *Ink-Jet-Based Rapid Prototyping System Introduced*, 4(3), CAD/CAM Publishing Inc., March, 1996: 6–7.

[20] Solidscape Inc., *ModelMaker II Case Studies: Design and Development of Fine Watches Time Piece Product* [Online]. Available: http://www.solid-scape.com/mmii_case_grotell.html, 2001.

[21]   Tsinghua University, *CLRF, Product Information* [Online]. Available: http://www.geocities.com/CollegePark/Lab/8600/clrfprod.htm, 2001.

[22]   Ennex Corporation, *All About Fabbers* [Online]. Available: http://www.ennex.com/fabbers/index.sht, 2001.

[23]   Burns, M., Hayworth, K.J., and Thomas, C.L., "Offset® Fabbing," Paper Presented at *Solid Freeform Fabrication Symposium*, University of Texas at Austin, USA, 1996.

[24]   Merz, R., Prinz, F.B., Ramaswami, K., Terk, M., and Weiss, L.E., "Shape Deposition Manufacturing," Paper Presented at *Solid Freeform Fabrication Symposium*, University of Texas at Austin, USA, 1994.

[25]   *The Shape Deposition Manufacturing Process: Methodology.* Available: http://www-2.cs.cmu.edu/~sdm/methodology.htm, 2001.

[26]   *The Shape Deposition Manufacturing Process.* Available: http://www-2.cs.cmu.edu/~sdm, 2001.

PROBLEMS

1. Describe the process flow of Cubic's Laminated Object Manufacturing.

2. Describe the process flow of Stratasys' Fused Deposition Modeling.

3. Compare and contrast the laser-based LOM™ process and the FDM systems. What are the advantages (and disadvantages) for each of the systems?

4. Describe the critical factors that will influence the performance and functions of:
    (i)   Cubic's LOM™,
    (ii)   Stratasys FDM,
    (iii) Kira Corporation's PLT,
    (iv)   3D Systems' MJM,
    (v)   Solidscape's Modelmaker.

5. Compare and contrast 3D Systems' MJM System with Solidscape's Modelmaker. What are the advantages (and disadvantages) for each of the systems?

6. Compare and contrast 3D Systems' MJM machines with its own liquid-based SLA® machines. What are the advantages (and disadvantages) for each of the systems?

7. Compare and contrast Cubic's LOM™ process with Kira Corporation's PLT and Ennex Corporation's Offset™ Fabber. What are the advantages (and disadvantages) for each of the systems?

8. What are the advantages and disadvantages of solid-based systems compared with liquid-based systems?

9. In the LOM™ systems, what do you think are the factors that limit the work volume of the systems?

10. What are the major advantages (or disadvantages) of hybrid RP system like that of Beijing Yinhua's Multi-functional RP system?

Chapter 5
POWDER-BASED RAPID PROTOTYPING SYSTEMS

5.1   3D SYSTEMS' SELECTIVE LASER SINTERING (SLS)

5.1.1   Company

3D Systems Corporation was founded by Charles W. Hull and Raymond S. Freed in 1986. The founding company, DTM Corporation, was established in 1987 to commercialize the SLS® technology. With the financial support from the BFGoodrich Company, and based on the technology that was developed and patented at the University of Texas at Austin, the company shipped its first commercial machine in 1992. DTM had worldwide exclusive license to commercialize the SLS® technology until they were bought over by 3D Systems in August 2001. 3D Systems' head office address is 26081 Avenue Hall, Valencia, CA91355, USA.

5.1.2   Products

5.1.2.1   *Model and Specifications*

In the last decade, the SLS® system has gone through three generations of products: the Sinterstation 2000, Sinterstation 2500 and the Sinterstation 2500^{plus} (see Figure 5.1). The latest and fourth generation SLS® system is the Vanguard™. The system is capable of producing objects measuring up to 380 mm (15 inches) length by 330 mm (13 inches) width by 380 mm (15 inches) in height, accommodating most rapid prototyping applications. The new Vanguard™ system offers several significant improvements over the previous generation systems

Figure 5.1: 3D Systems' SLS® system
(Courtesy 3D Systems)

such as improved part accuracy, higher speed, smoother surface finish and finer resolution. A summary of the specifications for the Vanguard™ si2™ is found in Table 5.1. The SLS® process is the only technology with the capability to directly process a variety of engineering thermoplastic materials, metallic materials, ceramic materials, and thermoplastic composites.

5.1.2.2   *Advantages*

(1) *Good part stability.* Parts are created within a precise controlled environment. The process and materials provide for directly produced functional parts to be built.

(2) *Wide range of processing materials.* A wide range of materials including nylon, polycarbonates, metals and ceramics are available, thus providing flexibility and a wide scope of functional applications.

(3) *No part supports required.* The system does not require CAD-developed support structures. This saves the time required for support structure building and removal.

(4) *Little post-processing required.* The finishing of the part is reasonably fine and requires only minimal post-processing such as particle blasting and sanding.

Table 5.1: Summary specifications of 3D Systems' Vanguard™ si2™ SLS® system

Model	Vanguard™ si2™ SLS®
Process	Selective Laser Sintering
Laser type	CO_2
Laser power (W)	25 or 100
Spot size (mm)	0.47
Maximum scan speed (mm/s)	7500 (standard beam delivery system) 10 000 (Celerity™ BDS)
XY resolution (mm)	0.178
Work volume, *XYZ* (mm × mm × mm)	370 × 320 × 445
Minimum layer thickness (mm)	0.076
Size of unit, *XYZ* (m × m × m)	2.1 × 1.3 × 1.9
Layering time per layer (s)	10 s
Data control unit	933 MHz Pentium III Windows 2000 OS
Power supply	240 V_{AC}, 12.5 kVA, 50/60 Hz, 3-phase

(5) *No post-curing required.* The completed laser sintered part is generally solid enough and does not require further curing.

(6) *Advanced software support.* The New Version 2.0 software uses a Windows® NT-style graphical user interface (GUI). Apart from the basic features, it allows for streamlined parts scaling, advanced nonlinear parts scaling, in-progress part changes, build report utilities and is available in foreign languages [1].

5.1.2.3 *Disadvantages*

(1) *Large physical size of the unit.* The system requires a relatively large space to house it. Apart from this, additional storage space is required to house the inert gas tanks used for each build.

(2) *High power consumption.* The system requires high power consumption due to the high wattage of the laser required to sinter the powder particles together.

(3) *Poor surface finish.* The as-produced parts tend to have poorer surface finish due to the relatively large particle sizes of the powders used.

5.1.3   Process

5.1.3.1   *The SLS® Process*

The SLS® process creates three-dimensional objects, layer by layer, from CAD-data generated in a CAD software using powdered materials with heat generated by a CO_2 laser within the Vanguard™ system. CAD data files in the STL file format are first transferred to the Vanguard™ system where they are sliced. From this point, the SLS® process (see Figure 5.2) starts and operates as follows:

(1) A thin layer of heat-fusible powder is deposited onto the part-building chamber.

(2) The bottom-most cross-sectional slice of the CAD part under fabrication is selectively "drawn" (or scanned) on the layer of powder by a heat-generating CO_2 laser. The interaction of the laser beam with the powder elevates the temperature to the point of melting, fusing the powder particles to form a solid mass. The intensity of the laser beam is modulated to melt the powder only in areas defined by the part's geometry. Surrounding powder remain a loose compact and serve as supports.

(3) When the cross-section is completely drawn, an additional layer of powder is deposited via a roller mechanism on top of the previously scanned layer. This prepares the next layer for scanning.

(4) Steps 2 and 3 are repeated, with each layer fusing to the layer below it. Successive layers of powder are deposited and the process is repeated until the part is completed.

As SLS® materials are in powdered form, the powder not melted or fused during processing serves as a customized, built-in support

Figure 5.2: The Selective Laser Sintering (SLS®) process
(Courtesy 3D Systems)

structure. There is no need to create support structures within the CAD design prior to or during processing and thus no support structure to remove when the part is completed.

After the SLS® process, the part is removed from the build chamber and the loose powder simply falls away. SLS® parts may then require some post-processing or secondary finishing, such as sanding, lacquering and painting, depending upon the application of the prototype built.

The SLS® system contains the following hardware components:

(1) Build chamber dimensions (381 × 330 × 457 mm)
(2) Process station (2100 × 1300 × 1900 mm)
(3) Computer cabinet (600 × 600 × 1828 mm)
(4) Chiller (500 × 800 × 900 mm)

The software that comes with the Vanguard™ si2™ SLS® System includes the Windows 2000 operating system and other proprietary application software such as the slicing module, automatic part distribution module, and part modification application software.

5.1.3.2 *Materials*

In theory, a wide range of thermoplastics, composites, metals and ceramics can be used in this process, thus providing an extensive range of functional parts to be built. The main types of materials used in the Vanguard™ si2™ SLS® System are safe and non-toxic, easy to use, and can be easily stored, recycled, and disposed off. These are as follows [2]:

- *Polyamide.* Trade named "DuraForm™", this material is used to create rigid and rugged plastic parts for functional engineering environments. This material is durable, can be machined or even welded where required. A variation of this material is the polyamide-based composite system, incorporating glass-filled powders, to produce even more rugged engineering parts. This composite material improves the resistance to heat and chemicals.
- *Thermoplastic elastomer.* Flexible, rubber-like parts can be prototyped using the SLS. Trade named, "SOMOS® 201", the material produces parts with high elongation. Yet, it is able to resist abrasion and provides good part stability. The material is impermeable to water and ideal for sports shoe applications and engineering seals.
- *Polycarbonate.* An industry-standard engineering thermoplastic. These are suitable for creating concept and functional models and prototypes, investment casting patterns for metal prototypes and cast tooling (with the RapidCasting™ process), masters for duplication processes, and sand casting patterns. These materials only require a 10–20 W laser to work and are useful for visualizing parts and working prototypes that do not carry heavy loads. These parts can be built quickly and are excellent for prototypes and patterns with fine features.
- *Nylon.* Another industry-standard engineering thermoplastic. This material is suitable for creating models and prototypes that can withstand and perform in demanding environment. It is one of the most durable rapid prototyping materials currently available in the industry, and it offers substantial heat and chemical resistance. A variation of this is the *Fine Nylon* and is used to create fine-featured parts for working prototypes. It is durable, resistant to heat and chemicals, and is excellent when fine detail is required.
- *Metal.* This is a material where polymer coated stainless steel powder is infiltrated with bronze. Trade named "LaserForm ST-100", the material is excellent for producing core inserts and pre-production tools for injection molding prototype polymer parts. The material exhibits high durability and thermal conductivity and can be used for relatively large-scale production tools. An alternative

material is the copper polyamide metal–polymer composite system which can be applied to tooling for injection molding small batch production of plastic parts.

- *Ceramics.* Trade named "SandForm™ Zr" and "Sandform™ Si", these use zircon and silica coated with phenolic binder to produce complex sand cores and molds for prototype sand castings of metal parts.

5.1.4 Principle

The SLS® process is based on the following two principles:

(1) Parts are built by sintering when a CO_2 laser beam hits a thin layer of powdered material. The interaction of the laser beam with the powder raises the temperature to the point of melting, resulting in particle bonding, fusing the particles to themselves and the previous layer to form a solid.
(2) The building of the part is done layer by layer. Each layer of the building process contains the cross-sections of one or many parts. The next layer is then built directly on top of the sintered layer after an additional layer of powder is deposited via a roller mechanism on top of the previously formed layer.

The packing density of particles during sintering affects the part density. In studies of particle packing with uniform sized particles [3] and particles used in commercial sinter bonding [4], packing densities were found to range typically from 50% to 62%. Generally, the higher the packing density, the better would be the expected mechanical properties. However, it must be noted that scan pattern and exposure parameters are also the major factors in determining the mechanical properties of the part.

5.1.4.1 *Sinter Bonding*

In the process, particles in each successive layer are fused to each other and to the previous layer by raising their temperature with the laser beam to above the glass-transition temperature. The glass-transition temperature is the temperature at which the material begins to soften from a solid to a jelly-like condition. This often occurs just prior to the melting temperature at which the material will be in a molten or liquid state. As a result, the particles begin to soften and deform owing to their weight and cause the surfaces in contact with other particles or solid to deform and fuse together at these contact surfaces. One major advantage of sintering over melting and fusing is that it joins powder particles into a solid part without going into the liquid phase, thus avoiding the distortions caused by the flow of molten material during fusing. After cooling, the powder particles are connected in a matrix that has approximately the density of the particle material.

As the sintering process requires the machine to bring the temperature of the particles to the glass-transition temperature, the amount of energy needed is considerable. The energy required to sinter bond a similar layer thickness of material is approximately between 300 to 500 times higher than that required for photopolymerization [5, 6]. This high laser power requirement can be reduced by using auxiliary heaters at the powder bed to raise the powder temperature to just below the sintering temperature during the sintering process. However an inert gas environment is needed to prevent oxidation or explosion of the fine powder particles. Cooling is also necessary for the chamber gas.

The parameters which affect the performance and functionalities are the properties of powdered materials and its mechanical properties after sintering, the accuracy of the laser beam, the scanning pattern, the exposure parameters and the resolution of the machine.

5.1.5   Applications

The Vanguard™ si2™ SLS® system can produce a wide range of parts in a broad variety of applications, including the following:

(1) *Concept models*. Physical representations of designs used to review design ideas, form and style.
(2) *Functional models and working prototypes*. Parts that can withstand limited functional testing, or fit and operate within an assembly.

(3) *Polycarbonate (RapidCasting™) patterns.* Patterns produced using polycarbonate, then cast in the metal of choice through the standard investment casting process. These build faster than wax patterns and are ideally suited for designs with thin walls and fine features. These patterns are also durable and heat resistant.

(4) *Metal tools (RapidTool™).* Direct rapid prototype of tools of molds for small or short production runs.

5.1.6 Examples

5.1.6.1 *Boeing Uses Prototyping to Maximize Return on Investment*

The Boeing Company, Air Vehicle Design Division, USA, had a mission to produce air vehicle configuration and integration designs [7]. Faced with the challenge to avoid tooling and production costs, reduce cycle time, increase efficiency in small batch production and maximize return on investment (ROI), Boeing used the Sinterstation® to prototyped visualization and technical review models, reproduce existing parts, and produced scaled models for laboratory testing. Boeing had also used DuraForm™ parts directly onto prototype aircraft, vehicles and mock-ups. The parts produced often undergo tests such as fatigue, strength, heat resistance and resistance to moisture. Figure 5.3 shows such a part prototyped for the aerospace and defense industry.

5.1.6.2 *Reebok Uses SLS Process for Developing Sports Shoes*

In the initial developmental stages of a new spikeless golf shoe sole design, Reebok's Golf Division, USA, needed a rapid, cost efficient way to create flexible prototypes for their design tests [8]. The company was up against very tight deadlines. Traditional prototyping methods (standard tooling and injection molding) would have taken 30 to 60 days and cost the company US$3500 to $4000 per prototype. The company decided to use the SLS® system to build SOMOS® 201 parts. This took only seven hours and about US$250 worth of materials. The prototype soles were affixed to a pair of golf shoes (see Figure 5.4) and

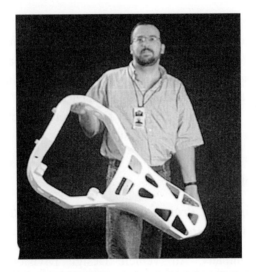

Figure 5.3: RP part for aerospace and defense industry
(Coutesy 3D Systems)

Figure 5.4: Prototypes for Reebok golf shoe soles produced by SLS®
(Courtesy 3D Systems)

Figure 5.5: Metal injection molding tools produced by SLS®
(Courtesy 3D Systems)

worn by an experienced golfer for two rounds of golf. The company
was able to save thousands of dollars and more than a month in
development time.

5.1.6.3   *Rover Applies SLS Process in Tooling for Injection Molding*

Rover Group Ltd. at Warwick, UK, produces automotive vehicles and
wanted to produce tooling for injection molding of nylon automobile
glass guides measuring 90 mm × 60 mm × 25 mm per part. The single
cavity tool insert was designed by CAD and then used to build the
hollow tool insert using the SLS system. The part was thereafter backed
with epoxy resin loaded with aluminium powder and granules before
machining it to fit a standard injection molding bolster set. Using this,
it shot 33 polypropylene and 117 nylon parts and applied several of
these on prototype vehicles. The Copper PA tool withstood an injection
molding temperature of 285°C. The company went on to create Copper
PA tooling for injection molding 300 speaker covers, each featuring
complex shapes, measuring a diameter of 40 mm, and which were made
of polypropylene. This resulted in significant time and cost savings,
providing Rover with a chance to do more design iterations to improve
the design without adding to the planned time frame and costs [9].

5.1.7 Research and Development

The primary research focus is on the development of new and advanced materials for fresh applications such as the medical industry. One such material is calcium hydroxyapatite, a material very similar to the human bone. 3D Systems is also continually working on improving and refining its SLS® process, software and system.

5.2 EOS's EOSINT SYSTEMS

5.2.1 Company

The EOSINT P, introduced in 1994, was the first European laser-sintering system for plastics as a logical complement to STEREOS by EOS GmbH Electro Optical Systems. The company, founded in 1989, is also the first European manufacturer of a laser-sintering-system. In 1995, the company launched the EOSINT M 250, the first worldwide system for direct laser-sintering and the EOSINT S 700, for direct manufacture of casting molds in foundry sand. In 1998, the EOSINT M 250 *Xtended* for direct laser-sintering of steel-based powder was launched. The next year, the EOSINT P 360 with increased speed and build volume was launched. In 2000, the company presented its first twin laser-sintering system, the EOSINT P 700, for plastics. Its new address, as from 16 September 2002, is at EOS GmbH Electro Optical Systems, Robert-Stiring-Ring 1, D-82152 Krailling, Germany.

5.2.2 Products

5.2.2.1 *Models and Specifications*

EOSINT systems share identical software, have similar optics and controller with EOS's STEREOS systems, but offer a different spectrum of building materials. The first system, EOSINT P, was designed for thermoplastic materials such as polystyrene and nylon. The other system, called EOSINT M, has been developed for direct sintering of metal powder. Table 5.2 shows the specifications of the EOS rapid prototyping systems.

Table 5.2: Specifications of EOSINT machines (Courtesy EOS GmbH)

Model	EOSINT P 360	EOSINT M250 *Xtended*	EOSINT S	EOSINT P 700
Laser type	CO_2	CO_2	CO_2	CO_2
Laser power (W)	50	Min. 200	2×50	50
XY sweep speed (m/s)	5	3	5 each	5
XY position accuracy (mm)	±0.05	±0.05	±0.05	±0.05
Work volume, XYZ (mm × mm × mm)	340 × 340 × 620	250 × 250 × 185	700 × 380 × 380	700 × 380 × 580
Layer thickness (mm)	0.1–0.2	0.05–0.1	0.2	0.15
Size of unit, XYZ (m × m × m)	2.15 × 1.3 × 1.25	1.95 × 1.85 × 1.1	2.1 × 1.4 × 1.4	2.1 × 1.41 × 2.27
Data control unit	PC Pentium Win 95, Win NT	PC Pentium Win 95, Win NT	PC Pentium Win 95, Win NT	PC Pentium Win 95, Win NT
Power supply	400 V_{AC}, 32 A	400 V_{AC}, 32 A	400 V_{AC}, 32 A	400 V_{AC}, 32 A

The EOSINT P 360 machine has a CO_2 laser and builds using thermoplastic powders. Two materials are presently available, polystyrene and nylon (polyamide). The principal applications of these materials are primarily for investment casting patterns and functional prototypes, but they can also be used for building geometrical models for visualization or form and fit testing.

EOSINT P 700 was presented in November 2000 and commercially available in April 2001. It is the first double-laser system for plastic laser-sintering and offers its users new dimensions such as productivity, building speed, part quality and also its range of applications. EOSINT P 700 builds fully functional plastic parts and investment casting patterns of any complexity layer by layer, directly on the basis of CAD data and in a single process. Being a modular system, it allows new technical features resulting from future development to be retrofitted

Figure 5.6: EOSINT P 700 machine (Courtesy of EOS GmbH)

in the future, thereby offering long term improvement protection. Figure 5.6 shows the EOSINT P 700 machine.

EOSINT M (see Figure 5.7) was introduced in late 1994 as a second stage of EOSINT development. It was the world's first commercial system for direct laser-sintering of metal powder. The material currently used is a special alloy mixture comprising mainly of bronze and nickel, developed by Electrolux Rapid Prototyping and licensed exclusively to EOS. This metal can be sintered without pre-heating and exhibits negligible net shrinkage during the sintering process. The DirectMetal™ 50-V2 is a fine-grained steel-based metal powder with a maximum particle size of 50 microns used for DirectTool™ applications on the EOSINT M laser-sintering system for very high precision, good detail resolution and smooth finishing.

EOSINT S was introduced to produce molds and cores for foundry sand metal casting in several hours from CAD data. The Direct Cast™ method requires no patterns or core boxes, commonly associated with sand casting, to rapidly produce extremely complex molds using layer by layer manufacturing technique with Direct Croning Process (DCP®). The process uses several coated foundry sands and offers an extremely

Figure 5.7: EOSINT M 250*Xtended* machine (Courtesy EOS GmbH)

Figure 5.8: EOSINT S 700 machine (Courtesy EOS GmbH)

fast route to sand mold production for metal prototypes. This is particularly cost effective for one-off or small batch component production. Figure 5.8 shows the EOSINT S 700 model which was launched in the year 2000.

5.2.2.2 *Advantages*

(1) *Good part stability.* Parts are created in a precise controlled environment. The process and materials provide for directly produced functional parts to be built.

(2) *Wide range of processing materials.* A wide range of materials including polyamide, glass-filled polyamide composite, polystyrene, metals and foundry sands thus providing flexibility and a wide scope of functional applications.

(3) *Support structures not required.* The EOSINT RP methods do not require support structures or uses only simplified support structures as in the case of the Direct Croning Process®. This increases the efficiency of the system by reducing the processing time of the build.

(4) *Little post-processing required.* The finishing of the part is very good thus requiring only minimal post-processing.

(5) *High accuracy.* For EOSINT P system, the polystyrene it uses can be laser-sintered at a relatively low temperature, thereby causing low shrinkage and high inherent building accuracy.

(6) *No post-curing required.* The completed part is by itself solid enough and does not require post-curing.

(7) *Large parts can be built.* The large build volume allows for relatively larger and taller parts to be built. Large single parts can be built at one go rather than by the building of smaller parts to be later joined together.

5.2.2.3   *Disadvantages*

(1) *Dedicated systems.* Only dedicated systems for plastic, metal and sand are available respectively.

(2) *High power consumption.* The EOSINT systems, especially the M model, require relatively high laser power in order to directly sinter the metal powders.

(3) *Large physical size of the unit.* The system requires a relatively large space to house.

5.2.3   **Process**

The EOSINT process applies the following steps to creating of parts, processing of data, preparation of the new layer, scanning and removal of unsintered powder [10]:

(1) First, the part is created in a CAD system on a workstation. Then, the CAD data are processed by EOS's software EOSOFT and converted to the cross-section format that EOSINT machines use to control the sintering process.

(2) At the build stage, a new powder layer covers the platform. The laser scans the new powder layer and sinters the powder together according to the cross-sectional data. Simultaneously, the new layer is joined to the previous layer.

(3) When the sintering of the cross-section is completed, the elevator lowers and another new layer is prepared for the next step. The processes are repeated till finally, the part is finished.

(4) After this, the powder around the part is removed.

The EOSINT system typically contains a Silicon Graphics workstation and an EOSINT machine including the working platform, a laser, and an optical scanner calibration system.

5.2.3.1   *Materials*

(1) *Standard Polyamides PA 1500, PA 1300 and PA 2200*. These are standard polyamides that are economical and recommended for new users and in cases of extreme time and cost pressure. These polyamides have larger grain sizes and are more ideal for building large prototypes with simpler geometries and less critical surface requirements. The fine PA 2200 powder can be used to offer both high functional performance and a sophisticated appearance. It has a smaller particle size and modified composition to allow for minute details of 0.55 mm resolution to be reproduced. The PA 2200 also allows for parts to withstand high-temperature painting.

(2) *Glass-filled Fine Polyamide PA 3200 GF*. The polymer composite is ideal for building load-bearing functional prototypes directly from 3D CAD data rapidly and economically. Parts made of glass-filled and fine-grained polyamide PA 3200 GF give extremely high accuracy and smooth surface finishes.

(3) *Polystyrene PS 1500 and PS 2500*. The PS 2500 polystyrene powder for use in the EOSINT P systems allows "dust-free

handling" patterns for use in investment casting and evaporative pattern casting methods to produce metal parts rapidly. The PS 2500 is an upgrade to the PS 1500 in terms of lower thermal expansion, distortion and residual ash content (0.1%).

(4) *DirectMetal*™ *50-V2*. This is a fine-grained metal powder with a maximum particle size and typical layer thickness of 50 microns. It is used for DirectTool™ applications on EOSINT M laser sintering systems and offers excellent mechanical properties, very high precision, good detail resolution and good surface finishing. This is primarily used for very complex injection molding tools.

(5) *DirectMetal*™ *100-V3*. This material has a maximum particle size of 100 microns and is suitable for high-speed tooling, such as laser-sintering injection molding and vulcanizing tools, for producing functional prototypes of plastic and rubber parts in production material. It offers good mechanical properties, high precision and detail resolution at very high building speeds.

(6) *Metal Powder M Cu 3201*. This is a copper-based metal powder of 50 microns that allows intricate structures and details to be produced using the Direct Metal Laser-Sintering (DMLS) technique. The material gives smooth surfaces with good electrical conductivity, thermal and mechanical properties.

(7) *DirectSteel*™ *50-V1*. Introduced in early 1999, this is a steel-based powder for application in injection molding, die casting and rapid part manufacturing. The material is a fine-grained powder with maximum particle size and typical layer thickness of 50 microns.

(8) *Foundry Sands*. EOSINT Zircon HT, made available in early 1998, is a high temperature casting zircon sand coated with phenolic resin allows for sand molds and cores to be built using the EOSINT S 700 system. The material has a low thermal expansion and high heat capacity and conductivity, making it ideal for high quality and high precision part castings. The EOSINT Quartz 4.2 is applied to producing "boxless" sand cores and molds. This material has very good sand core debonding properties and is suitable for very intricate geometries such as engine parts.

5.2.4 Principle

The principle of the EOSINT systems is based on the laser-sintering principle and layer manufacture principle similar to that of the SLS® described in Section 5.1.4.

The parameters that influence the performances and functionalities of the EOSINT systems are the properties of powder materials, the laser and the optical scanning system, the precision of the working platform and the working temperature.

5.2.5 Applications

(1) *Concept models.* Physical representations of designs used to visualize design ideas, form and style.
(2) *Functional models and working prototypes.* Parts that can withstand limited functional testing or fit and operate within an assembly. The system is suitable for the automotive, aerospace, machine tools and consumer products industries.
(3) *Wax and styrene cast patterns.* Patterns produced in wax, then cast in the metal of choice using the investment casting process. Styrene patterns, on the other hand, can evaporate pattern casted.
(4) *Metal tools.* The main application of the EOSINT system is in rapid tooling [11]. It is used primarily for creating tools for investment casting, injection molding and other similar manufacturing processes. Figure 5.9 shows a metal injection tool built by the EOSINT M250 *Xtended* for the purpose of the injection molding of plastic parts.

5.2.6 Examples

5.2.6.1 *FKM Sintertechnik Uses EOSINT P 700 to Build Fuel Tank Prototype*

FKM Sintertechnik GmbH of Biedenkopf-Breidenstein belongs to the first owners and users of EOSINT P700 pre-series system. In a project for an automotive manufacturer, FKM was asked to conduct thorough

Figure 5.9: Rapid metal tooling for injection molding of plastic parts
(Courtesy of EOS GmbH)

Figure 5.10: Fuel tank prototype produced on the EOSINT P 700
(Courtesy of EOS GmbH)

functional testing. At the same time, they wanted to verify the CAD data and check the design of the later series tooling. Since the tank had to be filled with an aggressive liquid, the customer paid great attention to the material's chemical properties. FKM therefore built the entire fuel tank in one piece on their EOSINT P 700 directly from polyamide and was able to deliver it within only four days. As an alternative, the tank could have produced with the help of a laminating tool. In this case, the customer would have had to wait two weeks for the first functional prototype [12].

5.2.7 Research and Development

EOS is working on more thermoplastic materials and post-processing and secondary processes for new applications and a still wider choice of materials for its EOSINT P machines. The same can be said for EOSINT M, for its laser-sintering materials in expansion of its application and wider choice of materials. The EOSINT S, which builds molds and cores for sand casting directly in foundry sand, has undergone beta-phase testing and become commercially available recently.

5.3 Z CORPORATION'S THREE-DIMENSIONAL PRINTING (3DP)

5.3.1 Company

Z Corporation was incorporated in 1994 by Hatsopoulos, Walter Bornhost, Tim Anderson and Jim Brett. It commercialized its first 3D Printer, the $Z^{TM}402$ System, based on three-dimensional technology (3DP) in 1997. This core technology was invented and patented at the Massachusetts Institute of Technology. It was subsequently licensed and further developed by Z Corporation. Z Corporation's distributors are headquartered in Australia, the Benelux area, France, Germany, Hong Kong, Italy, Japan, Korea, Malaysia, Russia, Singapore, South Africa, Spain, Taiwan, and United Kingdom. Its address is 20 North Avenue Burlington, MA 01803, United States of America.

5.3.2 Products

5.3.2.1 Models and Specifications

Z Corporation's latest products are the $Z^{TM}400$, $Z^{TM}406$ and $Z^{TM}810$ systems. The $Z^{TM}400$ System replaces the $Z^{TM}402$ System and has the same speed and performance as the $Z^{TM}402$ system but is configured for the entry-level and educational users. It does not come bundled with training or post-processing units. The Z406 3D Color Printer builds parts three to four times faster than Z402C and utilizes four print heads

Table 5.3: Specifications of Z Corporation's 3D Printers

Model	Z^{TM} 400 3DP	Z^{TM} 406 3DP	Z^{TM} 810 3DP
Build speed	2 layers / min	Color: 2 layers / min Monochrome: 6 layers / min	
Build volume (mm × mm × mm)	203 × 254 × 203	203 × 254 × 203	500 × 600 × 400
Layer thickness (mm)	0.076–0.254	0.076–0.254	0.076–0.254
Equipment dimensions (mm × mm × mm)	740 × 910 × 1070	740 × 910 ×1070	1020 × 790 ×1120
Equipment weight (kg)	136	136	210
System software	Z Corp.'s proprietary system software runs on Microsoft Windows 2000 and NT. VRML, ZCP, PLY and SFX file formats can be used for color input. STL file format is accepted for monochrome parts.		
Materials	Starch and plaster formulations.		

as compared to Z402C. Its new print heads were developed by Hewlett Packard. This new machine is the first product of a cross-licensing agreement between HP and Z Corporation in the field of 3D printing. Z^{TM}810 System's large build volume and inexpensive build materials make it the fastest and the least expensive way to create large appearance prototypes. The system also offers a variety of finishing options including epoxy infiltration, sanding, painting and plating. The option of color gives the user added information and aesthetics through the ability to incorporate color directly into the part as it is being printed [13]. Table 5.3 shows the specifications of Z Corporation's 3D printers.

The Z400 3D printer is the entry-level concept modeling solution that delivers great models quickly and inexpensively. Models can be used for design verification, communication and as patterns for casting applications. Z Corporation offers a variety of materials for use with the

Figure 5.11: Z™ 406 3D printer

Z400 3D printer. Beginning with two basic materials, a versatile and inexpensive starch-based powder and a high-definition plaster-based powder, infiltrants can also be added to satisfy a wide range of modeling needs.

The Z406 System is a premium 3D Printer with the capability of printing in full-color, communicating important information about parts, including engineering data, labeling, highlighting and appearance simulation. It can print in six million colors and uses a new pigment system developed by Cabot Corporation for fuller and brighter colors. The software interface included with the new machine, MAGICS Z allows users to add color information to STL files. It also includes a labeling option that lets users add text such as dates or revision coding directly to STL files. Figure 5.11 shows a photograph of Z Corporation's Z™ 406 3D printer.

The Z810 System is the fastest and the least expensive way to create large appearance prototypes for design review, mock-ups for form and fit testing, and patterns for casting applications. The large build volume allows full-scale concept models to be made for more effective communication with marketing, manufacturing, customers and suppliers. The Z810 System's color capability allows accurate representation of designs including FEA and other engineering data, further enhancing communication. Physical models can be created in plaster or starch-based materials and can be infiltrated to produce parts with a variety of material properties, satisfying a wide spectrum of modeling needs.

Z Corporation also has an accessory, the ZW4 Automated Waxer, for use with the 3D printers. The ZW4 Waxer allows printed parts to be infiltrated with paraffin wax to enhance strength, provide uniform part finish and color, or to create patterns suitable for investment casting.

5.3.2.2   *Advantages*

(1) *High speed.* Fastest 3D printer to date. Each layer is printed in seconds, reducing the prototyping time of a hand-held part to 1 to 2 hours.

(2) *Versatile.* Parts are currently used for the automotive, packaging, education, footwear, medical, aerospace and telecommunications industries. Parts are used in every step of the design process for communication, design review and limited functional testing. Parts can be infiltrated if necessary, offering the opportunity to produce parts with a variety of material properties to serve a range of modeling requirements.

(3) *Simple to operate.* The office compatible Zcorp system is straightforward to operate and does not require a designated technician to build a part. The system is based on the standard, off the shelf components developed for the ink-jet printer industry, resulting in a reliable and dependable 3D printer.

(4) *No wastage of materials.* Powder that is not printed during the cycle can be reused.

(5) *Color.* Enables complex color schemes in RP-ed parts from a full 24-bit palette of colors.

5.3.2.3   *Disadvantages*

(1) *Limited functional parts.* Relative to the SLS, parts built are much weaker, thereby limiting the functional testing capabilities.

(2) *Limited materials.* The materials available are only starch and plaster-based materials, with the added option to infiltrate wax using the ZW4 Waxer.

(3) *Poor surface finish*. Parts built by 3D printing have a relatively poorer surface finish and post-processing is frequently required.

5.3.3 Process [14]

(1) The machine spreads a layer of powder from the feed box to cover the surface of the build piston. The printer then prints binder solution onto the loose powder, forming the first cross-section. For monochrome parts, Z406 color printer uses all four print heads to print a single-colored binder. For multi-colored parts, each of the four print heads deposits a different color binder, mixing the four color binders to produce a spectrum of colors that can be applied to different regions of a part.

(2) The powder is glued together at where the binder is printed. The remaining powder remains loose and supports the layers that will be printed above.

(3) When the cross-section is completed, the build piston is lowered, a new layer of powder is spread over its surface, and the process is repeated. The part grows layer by layer in the build piston until the part is completed, completely surrounded and covered by loose powder. Finally the build piston is raised and the loose powder is vacuumed, revealing the complete part.

(4) Once a build is completed, the excess powder is vacuumed and the parts are lifted from the bed. Once removed, parts can be finished in a variety of ways to suit your needs. For a quick design review, parts can be left raw or "green." To quickly produce a more robust model, parts can be dipped in wax. For a robust model that can be sanded, finished and painted, the part can be infiltrated with a resin or urethane.

5.3.4 Examples

5.3.4.1 *Sports Shoe Industry*

The 3D printer has been used by designers, marketers, manufacturers, and managers in the footwear industry. Leading athletic shoe

Figure 5.12: Sports shoe design model created by Z Corporation system
(Courtesy of Z Corporation)

companies, such as Adidas, have used this RP system to radically reduce prototype development time and communicate in new ways [15]. Shoe industries these days are faced with constantly changing consumer preferences and have to react quickly to stay ahead of the business. With the 3D printer, lead times are drastically reduced, beating the competition to the shelves with the latest design trends whilst avoiding an excess inventory of unwanted designs.

5.3.4.2 *Javelin Puts Computer Sculpting in the Artist's Hands*

Javelin uses the Z402 System to produce computer-sculpted models to assist artists with complex modeling. Parts produced find applications with computer gamers, animators, pre-Hollywood mock-ups designers, and sculpturing artists. The low cost associated with 3DP parts allows several iterations to be used to accelerate the sculpting process. In one instance, CT scan data and Velocity2 software were used to recreate a dinosaur skull. They found that eliminating the need for support structure in a file that exceeds 4 000 000 polygons was a great asset. The CAD file from Velocity2 was sent to the Z402 System and the "dino-head" (shown in Figure 5.13) was built within 7 hours [16].

Figure 5.13: "Dino-head" produced by Javelin using the Z Corporation System
(Courtesy of Z Corporation)

5.3.5 Research and Development

On-going work to improve the versatility of the system in terms of materials, efficiency and software are being carried out to allow new applications.

5.4 OPTOMEC'S LASER ENGINEERED NET SHAPING (LENS)

5.4.1 Company

Optomec Inc. was incorporated in 1992. Since 1997, Optomec has focused on commercializing a direct fabrication process, the Laser Engineered Net Shaping (LENS™) process originally developed by Sandia National Laboratories. Optomec delivered its first commercial system to Ohio State University. The address of Optomec Inc. is 3911 Singer Boulevard, N.E., Albuquerque, NM 87109, USA.

5.4.2   Products

5.4.2.1   *Model and Specifications*

The latest Optomec's products are the LENS™ 750 and LENS™ 850 systems. These two systems feature the Laser Engineered Net Shaping (LENS) process, a technology that builds or repairs parts using metal powders to form fully dense objects to give excellent material properties. This technique can be used with a wide variety of metals including titanium, tool steels, stainless steels, copper and aluminum. The LENS™ 750 and LENS™ 850 systems contains the following hardware components as in Table 5.4.

Table 5.5 shows a summary of the models and specifications of the LENS systems. Figure 5.14 shows a photograph of Optomec's LENS™ 750 system.

5.4.2.2   *Advantages*

(1) *Superior material properties*. The LENS process is capable of producing fully dense metal parts [17]. Metal parts produced can also include embedded structures and superior material properties. The microstructure produced is also relatively good.
(2) *Complex parts*. Functional metal parts with complex features are the forte of the LENS system.
(3) *Reduced post-processing requirements*. Post-processing is minimized, thus reducing cycle time.

Table 5.4: Hardware components of the LENS systems

LENS™ 750	LENS™ 850
Argon recirculation unit	Argon recirculation unit
Laser power supply	Laser power supply
Ante-chamber	Ante-chamber
Workstation	Workstation
Process chamber and dri-train	Glove box

Table 5.5: Summary specifications of Optomec's LENS™ systems
(Courtesy of Optomec Inc.)

Model	LENS™ 750	LENS™ 850
Process	LENS	LENS
Build volume, *XYZ* (mm × mm × mm)	300 × 300 × 300	460 × 460 × 1070
Laser type	Nd:YAG single head laser	Nd:YAG dual head laser
Laser power (W)	600	1000
Laser wavelength (mm)	1064	1064
XY resolution (mm)	0.5	0.5
Z resolution (mm)	5	5
Size of unit, *XYZ* (mm × mm × mm)	1830 × 1040 × 2080	1170 × 1245 × 2080
Machine size (kg)	2540–2858	2540–2858
Power supply	208 or 240 V_{AC}, 3-phase, 75 A	460 V_{AC}, 3-phase, 75 A
Workstation	Pentium III Windows NT	Pentium III Windows NT
Materials	Stainless steels, H13 tool steel, titanium, super alloys such as inconel, tungsten, copper and aluminium	

Figure 5.14: Optomec's LENS™ 750 system
(Courtesy of Optomec Inc.)

5.4.2.3 *Disadvantages*

(1) *Limited materials.* The process is currently narrowly focused to produce only metal parts.
(2) *Large physical unit size.* The unit requires a relatively large area to house.
(3) *High power consumption.* The laser system requires very high wattage.

5.4.3 Process

The LENS™ process builds components in an additive manner from powdered metals using a Nd:YAG laser to fuse powder to a solid as shown in Figure 5.15. It is a freeform metal fabrication process in which a fully dense metal component is formed. The LENS™ process comprises of the following steps [18]:

(1) A deposition head supplies metal powder to the focus of a high powered Nd:YAG laser beam to be melted. This laser is typically directed by fiber optics or precision angled mirrors.

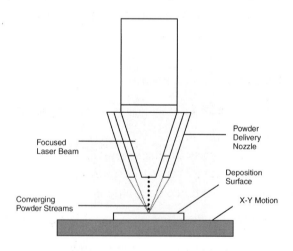

Figure 5.15: Optomec's LENS process

(2) The laser is focused on a particular spot by a series of lenses, and a motion system underneath the platform moves horizontally and laterally as the laser beam traces the cross-section of the part being produced. The fabrication process takes place in a low-pressure argon chamber for oxygen-free operation in the melting zone, ensuring that good adhesion is accomplished.

(3) When a layer is completed, the deposition head moves up and continues with the next layer. The process is repeated layer by layer until the part is completed. The entire process is usually enclosed to isolate the process from the atmosphere. Generally the prototypes need additional finishing, but are fully dense products with good grain formation.

5.4.4   Principle

The LENS™ process is based on the following two principles:

(1) A high powered Nd:YAG laser focused onto a metal substrate creates a molten puddle on the substrate surface. Powder is then injected into the molten puddle to increase material volume.

(2) A "printing" motion system moves a platform horizontally and laterally as the laser beam traces the cross-section of the part being produced. After formation of a layer of the part, the machine's powder delivery nozzle moves upwards prior to building next layer.

5.4.5   Applications

The LENS technology can be used in the following areas:

(1) Build mold and die inserts
(2) Producing titanium parts in racing industry
(3) Fabricate titanium components for biological implants
(4) Produce functionally gradient structures

Figure 5.16 shows a photograph of a metallic hip implant produced using LENS.

Figure 5.16: Medical LENS produced metal part
(Courtesy of Optomec Inc.)

5.4.6   Research and Development

The LENS™ Technology Group's future plans include continued research into embedded structures, thermally conductive materials, single crystal applications, gradient materials, metal matrix composites, mold repair and modification and ways to increase deposition rate.

Optomec is also developing software for a five-axis head, allowing the process to handle more difficult geometries. The new head will permit the deposition of metal in areas that have walls that are 90° to each other.

5.5   SOLIGEN'S DIRECT SHELL PRODUCTION CASTING (DSPC)

5.5.1   Company

Soligen Technologies Inc. was founded by Yehoram Uziel, its President and CEO, in 1991 and went public in 1993. It first installed its Direct Shell Production Casting (DSPC) System at three "alpha" sites in 1993. It bought the license to MIT's 3D printing patents for metal casting which is valid till 2006. Its address is Soligen Inc., 19408 Londelios St., Northridge, California 91324, USA.

Figure 5.17: Soligen's Direct Shell Production Casting System, DSPC 300
(Courtesy Soligen Inc.)

5.5.2 Product

5.5.2.1 *Model and Specifications*

Direct Shell Production Casting (DSPC) creates ceramic molds for metal parts with integral coves directly and automatically from CAD file. Soligen's Direct Shell Production Casting machine (see Figure 5.17), DSPC 300, includes the following mechanisms:

(1) A powder holder which contains the manufacturing material — powder.
(2) A powder distributor to distribute a thin layer of powder.
(3) Rollers which are used to compress each layer before binding.
(4) A print head which sprays binder on each layer.
(5) A bin which is used to hold the mold.

The specifications of the machine are summarized in Table 5.6.

5.5.2.2 *Advantages*

(1) *Patternless casting.* Direct tooling, thus eliminating the need to produce any patterns.

Table 5.6: Summarized specifications of DSPC 300
(Courtesy Soligen Inc.)

Model	DSPC 300
Process	Direct Shell Production Casting
XY resolution (mm)	0.05
Work volume, XYZ (mm × mm × mm)	304 × 304 × 304
Layer thickness (mm)	0.178
Vertical build rate (mm/hour)	12.7–19.0
Size of unit, XYZ (m × m × m), estimated	2300 × 1300 × 2600
Data control unit	EWS
Power supply	110 or 220 V_{AC}, 10 A

(2) *Functional metal parts*. Up till the late 1990's, DSPC was the only rapid prototyping process which created ceramic molds for metal casting. As a result, functional metal parts (or metal tooling, such as dies for die casting) could be made directly from the CAD data of the part.

(3) *Net-shaped integral molds*. No parting lines, core prints or draft angles are required. Integral gatings and chills can be added to optimize mechanical properties.

5.5.2.3   *Disadvantages*

(1) *Limited materials*. The DSPC 300 only focuses on making ceramics molds primarily for metal casting.

5.5.3   Process

The DSPC technology is derived from a process known as three-dimensional printing and was invented and developed at the Massachusetts Institute of Technology (MIT), USA. The process steps are illustrated in Figure 5.18 and comprises of the following steps [19, 20]:

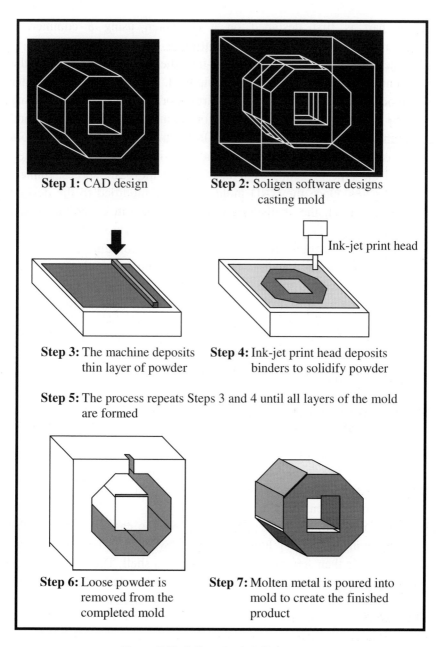

Step 1: CAD design

Step 2: Soligen software designs casting mold

Ink-jet print head

Step 3: The machine deposits thin layer of powder

Step 4: Ink-jet print head deposits binders to solidify powder

Step 5: The process repeats Steps 3 and 4 until all layers of the mold are formed

Step 6: Loose powder is removed from the completed mold

Step 7: Molten metal is poured into mold to create the finished product

Figure 5.18: Soligen Inc.'s DSPC process

(1) A part is first designed on a computer, using a commercial computer aided design (CAD) software.

(2) The CAD model is then loaded into the shell design unit — the central control unit of the equipment. Preparing the computer model for the casting mold requires modifications such as scaling the dimensions to compensate for shrinkage, adding fillets, and removing characteristics that will be machined later. The mold maker then decides how many mold cavities will be on each shell and the type of gating system, including the basic sprues, runners and gates. Once the CAD mold shells are modified to the desired configuration, the shell design unit generates an electronic model of the assembly in slices to the specified thickness. The electronic model is then transferred to the shell production unit.

(3) The shell production unit begins depositing a thin layer of fine alumina powder over the shell working surface for the first slice of the casting mold. A roller follows the powder, leveling the surface.

(4) An ink-jet print head, similar to those in computer printers, moves over the layer, injecting tiny drops of colloidal silica binder onto the powder surface from its 128 ink jets. Passing the pressurized stream of binder through a vibrating piezoelectric ceramic atomizes it as it exits the jet. The droplets pick up electric charge as they pass through an electric field which helps to align them to the powder. The binder solidifies the powder into ceramic on contact and the unbounded alumina remains as support for the following layer. The work area lowers and another layer of powder is distributed.

(5) The process through Steps 3 and 4 is repeated until all layers of the mold have been formed.

(6) After the building process is completed, the casting shell remains buried in a block of loose alumina powder. The unbound excess powder is then separated from the finished shell. The shell can then be removed for post-processing, which may include firing in a kiln to remove moisture or preheating to an appropriate temperature for casting.

(7) Molten metal can then be poured in to fill the casting shell or mold. After cooling, the shell can be broken up to remove the cast which can then be processed to remove gatings, sprues, etc., thus completing the casting process.

The hardware of the DSPC system contains a PC computer, a powder holder, a powder distributor, rolls, a print head and a bin. The software includes a CAD system and a Soligen's slicing software.

5.5.4   Principle

The principle of Soligen's Direct Shell Production Casting (DSPC) is based on Three-Dimensional Printing (3DP), a technology invented, developed and patented by Massachusetts Institute of Technology (MIT). 3DP is licensed exclusively to Soligen on a worldwide basis for the field of metal casting. Using this patented inkjet technology, binder from the nozzle selectively binds the ceramic particles together to create each layer. Layer after layer is built and bound to the previous layer until the ceramic shell is completed. The shell is removed from the DSPC machine and fired to vitrification temperatures to harden and remove all moisture. All excess ceramic particles are blown away.

In the process, the parameters that influence performance and functionality are the layer thickness, the powder's properties, the binders and the pressure of the rollers.

5.5.5   Applications

The DSPC Technology is used primarily to create casting shells for production of parts and prototypes. Soligen Inc. has acquired a foundry in Santa Ana for the commercial production of the prototypes and molds, called Parts Now™ Division. It serves as a service center as well as a place for equipment and process research and development. It aims to be a premier "one-stop shop" for functional cast metal parts produced directly from a CAD file, and with no need for pre-fabricated tooling to produce the first article. DSPC has been used in the following areas:

(1) Automotive industry.
(2) Aerospace industry.
(3) Computer manufacture.
(4) Medical prostheses.

5.5.6   Examples

5.5.6.1   *Soligen's Parts Now™ Pilots Metal Parts for Caterpillar*

Caterpillar needed to prototype a complex engine component within a week [21]. They sent the design file of the component via a modem to Parts Now™ (a division of Soligen Inc.) on Wednesday evening. By Friday, Parts Now™ had completed two casting shells with the DSPC systems. On Monday, the foundry poured the shells with A356 aluminum (see Figure 5.7). On Tuesday, the parts were heat treated and machining finished on Wednesday. On the same day, the fully functional parts were shipped to Caterpillar for installation on the engine, thus completing the order from Caterpillar. This is very much faster than any traditional fabrication process could deliver.

5.5.7   Research and Development

Soligen Inc. continues to do research on part- and mold-design. Using knowledge and experience from application development, the Parts Now™ division is developing technology to produce intricate aluminum

Figure 5.19: The cast aluminium engine component for Caterpillar by Parts Now™
(Courtesy of Soligen Inc.)

die casts that could improve throughput from 50% to 250% while improving accuracy and durability as well. Improving mold surface finishes with the DSPC-cast electrode used with electric-discharge machining (EDM) and coating technologies are also under development.

5.6 FRAUNHOFER'S MULTIPHASE JET SOLIDIFICATION (MJS)

5.6.1 Company

The Fraunhofer-Gessellschaft is Germany's leading organization of applied research, maintaining 46 research establishments at 31 locations [22]. Two of these establishments, the Fraunhofer Institute for Applied Materials Research (IFAM) and the Fraunhofer Institute for Manufacturing Engineering and Automation (IPA) cooperated in developing a RP process named multiphase jet solidification (MJS). The first machine was assembled and tested at IFAM (considered as an alpha site). The address for IFAM is Fraunhofer IFAM, Lesumer Heerstraße 36, D-28717, Bremen, Germany. The address for IPA is Fraunhofer IPA, Nobelstraße 12, D-70569, Stuttgart, Germany.

5.6.2 Process

The MJS process is one that is able to produce metallic or ceramic parts. It uses low-melting point alloys or a powder-binder mixture which is squeezed out through a computer-controlled nozzle to build the part layer by layer.

The MJS process comprises two main steps: data preparation and model building.

(1) *Data preparation.* In the first step, the part is designed on a 3D CAD system. The 3D CAD data is then imported to the MJS system, and together with process parameters like machining speed and materials flow rate, a controller file for the machine is then generated. This file is subsequently downloaded onto the controller of the machine for the build process.

(2) *Model building*. This is essentially the build process of the MJS process. The material used is usually a powder-binder-mixture but it can also be a liquefied alloy. At the beginning of the build process, the material is heated to beyond its solidification point in a heated chamber. It is then squeezed out through a computer-controlled nozzle by a pumping system, and deposited layer by layer onto a platform [23, 24]. The melted material solidifies when it comes into contact with the platform or the previous layer as both temperature and pressure decrease and heat is transferred to the part and the surrounding environment. The contact of the liquefied material leads to partial remelting of the previous layer and thus bonding between layers results.

After one cross-section is finished, the extrusion jet which is mounted on a *XYZ* table that is controlled by the system's computer moves in the *Z*-direction and next layer is built. The part is built layer by layer until it is completed.

The main components of the apparatus used for the MJS process comprises a personal computer, a computer-controlled positioning system and a heated chamber with a jet and a hauling system. The machine precision of the computer-controlled positioning system is ±0.01 mm and is able to traverse 500 mm × 540 mm × 175 mm on the *X*, *Y*, and *Z* axes (this is effectively the work volume). The chamber is temperature-stabilized and can be varied within ±1°C. The material is supplied as powder, pellets or bars. The extrusion temperature of the molten material can reach up to 200°C. Extrusion orifices vary from 0.5 to 2.0 mm.

5.6.3   Principle

The working principle of the MJS process [25] is shown in Figure 5.20. The basic concept applies the extrusion of low viscosity materials through a jet layer by layer, somewhat comparable to the fused deposition modeling process (see Section 4.2). The main differences between the two processes are in the raw material used to build the model and the feeding system. For the MJS process, the material is supplied in different phases using a power-binder-mixture or liquefied

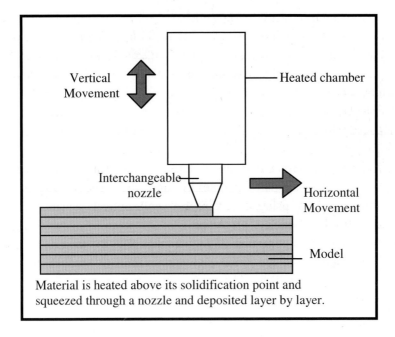

Figure 5.20: Working principle of the MJS process

alloys instead of using material in the wire-form. As the form of the material is different, the feed and nozzle system is also different. The material used is wax loaded with up to an approximately 50% volume fraction of metal powder.

In the MJS process, parameters that influence its performance and functionality are the layer thickness, the feed material, i.e., whether it is liquefied alloys (usually low melting-point metals) or powder-binder-mixture (usually materials with high melting point), the chamber pressure, the machining speed (build speed), the jet specification, the material flow, and the operating temperature.

5.6.4 Applications

The MJS technology can be used in many areas where functional metal parts are required, such as the automotive, aerospace, biomedical and machine tools industries.

The advantages of the MJS process are its capability to process a high variability of materials, e.g., high melting metals and ceramics, and the simplicity of its apparatus. To date, the system has been used to produce parts from silicon carbide, stainless steel 316L, titanium, alumina and bronze powder.

5.6.5   Research and Development

The focus of research activities at IFAM is to optimize the process so as to increase the part complexity and the process temperature. Research involves using touch probes and optical sensors and the metallic coating of plastic prototypes.

The melting temperatures of industrially relevant materials for the direct metal process, e.g., zinc and aluminum, are much higher than those of the tested alloys. Therefore, it is important to improve the apparatus to be able to handle these higher temperatures. The optimization of the materials' behavior for the process is also important in order to increase the complexity and the accuracy of the parts. Suitable powder combinations for other metals, e.g., copper and titanium, will be investigated and developed. Research will also be done to make possible the production of ceramic parts, such as alumina or zirconia through the MJS process.

5.7   ARCAM'S ELECTRON BEAM MELTING (EBM)

5.7.1   Company

Arcam AB, a Swedish technology development company was founded in 1997. The company's main activity is concentrated in the development of the Electron Beam Melting (EBM) technique for the production of solid metal parts directly from metal powder based on a 3D CAD model. The fundamental development work for Arcam's technology began in 1995 in collaboration with Chalmers University of Technology in Gotherburg, Sweden. The Arcam EBM technology was commercialized in 2001. The company address is at Krokslatts Fabriker 30, SE-431 37 Molndal, Sweden.

5.7.2 Products

The Electron Beam Melting (EBM) process is used to produce metal parts directly from a CAD model. The Arcam EBM S12 includes the following hardware components:

- Electron Beam gun with sweeping system,
- Vacuum chamber with fabrication tank and powder holder/setter,
- Vacuum pumps,
- Monitor,
- Linear device,
- High voltage unit,
- Electronic control system,
- Control unit.

5.7.2.1 *Model and Specifications*

Table 5.7 shows the specification for the Arcam EBM S12. Figure 5.21 shows a photograph of the Arcam EBM S12 system. Currently, only H13 tool steel material is used. An operating temperature of more than 1000°C is required to rapid prototype parts.

5.7.3 Process

(1) The part to be produced is first designed in a 3D CAD program. The model is then sliced into thin layers, approximately a tenth of a millimeter thick.
(2) An equally thin layer of powder is scraped onto a vertically adjustable surface. The first layer's geometry is then created through the layer of powder melting together at those points directed from the CAD file with a computer-controlled electron beam.
(3) Thereafter, the building surface is lowered and the next layer of powder is placed on top of the previous layer. The procedure is then repeated so that the object from the CAD model is shaped layer by layer until a finished metal part is completed.

Table 5.7: Specifications of the Arcam EBM S12

Model	ARCAM EBM S12
Dimension (mm)	$1800 \times 900 \times 2200$ (W × D × H)
Building volume (mm)	$250 \times 250 \times 200$
Melting speed (m/s)	0.5–1 (material dependent)
Layer thickness (mm)	0.05–0.5
Electron beam scan speed	Up to 1 km/s
Electron position accuracy (nm)	+50
Compressed air supply	4.5 bar, 0.1 m³/h
Power supply	3×400 V, 32 A
Process computer	PC, Windows NT
Software	Arcam tools, Magics RP/Materialise

Figure 5.21: Arcam's Electron Beam Melting (EBM) S12 system
(Courtesy of Arcam AB, Sweden)

5.7.4 Principle

The EBM process is based on the following two principles [26]:

(1) Parts are built up when an electron beam is fired at the metal powder. The computer controlled electron beam in vacuum melts the layer of powder precisely as indicated by the CAD model with the gain of the electrons' kinetic energy.
(2) The building of the part is accomplished layer by layer. A layer is added once the previous layer has melted. In this way, the solid details are built up of thin metal slices melted together.

5.7.5 Applications

The EBM process is used to manufacture H13 tool steel injection and compression molding tools, functional prototypes and components in small batches.

5.8 AEROMET CORPORATION'S LASFORM TECHNOLOGY

5.8.1 Company

In 1997, AeroMet™ was formed as a subsidiary of MTS Systems Corporation (MTS). The first Lasform™ system, currently installed at AeroMet™'s Eden Prairie, Minnesota facility, is being operated in collaboration with the U.S. Army Research Laboratory (ARL) of the Aberdeen Proving Ground, Maryland. The AeroMet™ Lasform™ technology is based on research performed jointly by the Applied Physics Laboratory of the John Hopkins University, the Applied Research Laboratory of Penn State University and the MTS Systems Corporation in 1996 and 1997. The company specializes in the laser additive manufacturing of high quality titanium alloys structures. Its address is 7623 Anagram Drive, Eden Prairie, Minnesota 55344, USA.

5.8.2 Products

The AeroMet™ laser forming machine (see Figure 5.22) is a three-axis system which can support the fabrication of parts within an inert processing chamber up to 3 m × 3 m × 1.2 m in size. The X and Y axes of motion are provided by a two-axis table which moves the part. The Z motion is achieved by vertically moving the coaxial laser-delivery/powder-nozzle assembly. The very large work volume chamber allows the processing of extremely large components in a protective atmosphere of argon containing less than 150 ppm of oxygen. The 18 kW CO_2 laser provides the power to obtain high deposition rates. The powder-feed system ensures a regular flow of powder at the high mass flow rates associated with this process.

The Lasform™ process uses commercially available precursor materials and creates parts that require minimal post-machining or heat treatment prior to use. As the precursor material is in the form of metal powders, it is also possible to produce "graded alloys" across the geometry of a component via real-time mixing of elemental

Figure 5.22: AeroMet™'s Lasform™ machine
(Courtesy of AeroMet Inc.)

constituents. This is a very unique feature of great interest to RP users and designers [27]. Since the AeroMet™ process takes place in an inert environment, it is possible to use the Lasform process in niobium, rhenium and other reactive materials which require protective processing atmosphere.

5.8.2.1 *Advantages*

(1) *High quality titanium parts*. Lasform^{SM} offers very high quality titanium parts to be rapidly produced. Parts mechanical tested reveal that AMS standards for commercially pure Ti, Ti-6Al-4V, and Ti-5Al-2.5Sn are met.
(2) *Very large parts*. Because of its very large work volume, very large parts can be seamlessly accommodated, whilst maintaining an oxygen-free, inert atmosphere.
(3) *Cost and time savings*. These laser-formed shapes require minimal post-machining and heat treating. This provides substantial cost and time savings by eliminating high materials waste, costly manufacturing tooling and long machining times.
(4) *Flexibility*. Lasform^{SM} offers flexibility through its ability to vary the composition of the material throughout the part that could result in the formation of a functionally gradient material where the microstructure and mechanical properties can vary depending on the composition. Thus the composition of different regions of the part can be customized according to functional and economic requirements.

5.8.2.2 *Disadvantages*

(1) *Very large physical unit size*. A very large space has to be dedicated to house this system.
(2) *Variety of material*. The system specializes only on the fabrication of titanium parts and metals. It is therefore cannot be used to produce polymer parts or parts to be used as non-functional models.

5.8.3   Process

Figure 5.23 illustrates the Lasform^{SM} process. The process is described as follows [28]:

(1) The AeroMet™ laser-forming process starts with a CAD representation of the part. This is then translated via proprietary software to generate trajectory paths for the laser-forming system. These paths are transmitted as machine instructions to the laser-forming system.

(2) The focused laser beam traces out the structural shape pattern of the desired part by moving the titanium target plate beneath the beam in the approximate $x-y$ trajectories.

(3) Titanium pre-alloyed powder is introduced into the molten metal head and provides for the build up of the desired shape as the molten spot is traversed over a target plate in the desired pattern.

(4) The 3D structure is fabricated by repeating the pattern, layer by layer over the desired geometry and indexing the focal point one layer up for the repeating pattern. This layer-by-layer registry with metallurgical integrity between layers generates the desired integral ribbed structure called a machining pre-form. Post-processes include heat treatment, machining and inspection.

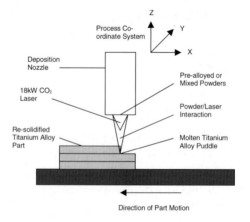

Figure 5.23: Schematic diagram of the Lasform^{SM} process

5.8.4 Principle

The Lasform[SM] process uses gas atomized and hydride–dehydride titanium alloy powders introduced into the focus region of the CO_2 laser beam [29]. The focus region is shifted in the X–Y plane as determined by the CAD slice. This is achieved by driving a numerical controlled manipulator to reproduce the desired shape. A solid titanium deposit layer remains and the process is repeated for the next layer in the Z direction. The new layer is also fused with the previous one, building layer upon layer until the part is completed. The process is carried out in an argon-filled environment. The production of high quality titanium shapes via laser direct metal deposition requires the integration of several technologies. These include high power laser beam generation and delivery, metal powder handling, robotics, process sensing and control, and environmental controls. AeroMet™ has been able to integrate these technologies and offer the Lasform[SM] on a commercial basis.

5.8.5 Applications

The Lasform[SM] process can be applied successfully in the manufacturing of integrally reinforced ribbed structural components for advanced aircraft [29]. Using the Lasform process, the desired geometry can be obtained simply by building the desired near net shape structure from fused titanium (alloy) powder. Applications in alloy systems other than titanium are also being identified. For example, the forging industry has a need to produce relatively large molds and dies quickly and inexpensively directly out of metal.

5.8.6 Research and Development

There is ongoing research into materials that can be used with the Lasform[SM] process as well as further improvements being made to the present Lasform[SM] machine.

5.9   GENERIS' RP SYSTEMS (GS)

5.9.1   Company

Generis is a German-based start-up company founded in May 1999 by Dr. Ingo Ederer, Prof. Heinzl and Rainer Hochsmann. Generis actively develops, produces and markets machinery for rapid prototyping. The GS 1500, Generis's first product centers on a process that builds molds for sand casting based on three-dimensional printing technology. Its address is Am Mittleren Moos 15, D-86167 Augsburg-Lechhausen, Germany.

5.9.2   Products

5.9.2.1   *Model and Specifications*

GS 1500 machine, as shown in Figure 5.24, makes basic as well as complex sand molds and sand cores for cast metal from CAD data. Standard casting-industry materials are used, enabling easy integration of the unit into existing processing procedures. The large building area and high building speed makes fast and tool-free production of metal-based prototypes and small-scale productions possible. A summary of the GS 1500 machine specifications is shown in Table 5.8.

5.9.3   Principle

The principle of the Generis process is based on Three-Dimensional Printing (3DP), a technology invented, developed and patented by Massachusetts Institute of Technology (MIT).

5.9.4   Applications

The Generis process can produce sand molds and cores for use in sand casting.

5.9.5   Research and Development

Generis GmbH and Soligen Technologies Inc. had reached an agreement in November 2000 to collaborate in the development and

Figure 5.24: The Generis GS 1500
(Courtesy of Generis)

Table 5.8: Specifications of GS 1500 machine
(Courtesy of Generis)

Model	GS 1500
Building volume	1500 mm × 750 mm × 750 mm (L × B × H)
Layer thickness	0.15–0.4 mm
Resolution	0.2 mm
Accuracy	0.1%
Dimensions	3100 mm × 3550 mm × 2500 mm (L × B × H)
Weight	2000 kg
Power connection	400 V, 3 × 16 A
Power consumption	5 KW
Software: Process control system	Pentium 3, 500 MHz
Operating system	WinNT 4.0
Data interface	CLI

marketing of equipment to create molds for metal casting directly from CAD data. Soligen, the worldwide exclusive licensee of MIT for 3DP patents in metal casting has sublicensed Generis the rights to develop, manufacture and market devices and methods for production of resin bound sand molds metal casting, made directly from CAD data [30]. Generis will appoint Soligen the exclusive distributor of Generis products in North America.

5.10   THERICS INC.'S THERIFORM TECHNOLOGY

5.10.1   Company

Therics Inc., a biopharmaceutical company, is a subsidiary of Tredeger Corporation. Based upon the three-dimensional printing technology developed at the Massachusetts Institute of Technology, the TheriForm™ technology is protected by a broad portfolio of strong patents, including an exclusive license from MIT for the worldwide health care market. From 1993, Therics has worked closely with MIT to develop successful generations of TheriForm™ fabrication machines that can create both macro-shapes and complex micro architectures that can meet the stringent regulatory requirements applicable to healthcare products. Its address is 115 Campus Drive, Princeton, NJ 08540, USA.

5.10.2   Products

The first fully automatic TheriForm™ machine, the Series 1100, was introduced in July 1997. The second generation TheriForm™ machine, the Series 2100, became operational in January 1998 with a build volume that is twice the size of Series 1100 machine. This Series 2100 machine, the company's first production scale machine, is suited for the commercial manufacturing of implantable drugs and tissue engineering products for both bone and soft tissue applications. The system is also appropriately sized for the production of clinical supplies of solid oral products and has been successfully used to produce numerous pharmaceutical products including tablets for stability studies.

In January 1999, Therics launched the Series 3200 TheriForm™ machine, the first high capacity solid oral tablet production equipment with an initial production rate of 20 000 oral dosage forms per hour. The 3200 machine can also be used for the commercial production of tissue engineering products.

5.10.3 Process

The TheriForm™ manufacturing process (see Figure 5.25) works in a manner similar to an ink jet printer, creating three-dimensional products composed of a series of two-dimensional layers. The process contains the following steps [31]:

(1) Products are fabricated by printing micro drops of binders, drugs and other materials and even living cells onto an ultra-thin layer of powdered polymers and biomaterials in a computer-directed sequence.

(2) These droplets bind with the powder to form a particular two-dimensional layer of the product. After each layer is printed, the build platform descends and a new layer of powder is spread and printing process is repeated.

(3) Thus the successive layers are built upon and bind to previous layer until the entire structure is completed.

5.10.4 Principle

TheriForm™ technology is based on the 3D printing technology developed by MIT. Binder droplets are selectively dispensed to bond the powders in the same layer together as well as to the next layer.

5.10.5 Applications

TheriForm™ technology is ideally suited for the creation of unique reconstructive implants including bone replacement products from three-dimensional imaging data (e.g., from MRI or CT) [32]. It is also

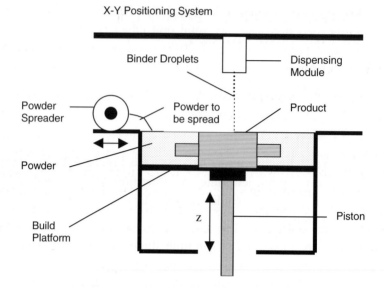

Figure 5.25: Schematic of the TheriForm™ process

used to produce cell-containing tissue that can incorporate drugs and growth factors at specific geometrical locations in functional gradients to enhance the development of functional tissue within the body [33]. TheriForm™'s ability to fabricate structures with architectures and material compositions provides a useful RP system that responds to the challenge of administrating drugs, proteins and biologicals. This is particularly important when conventional manufacturing methods do not provide a high level of precision and versatility necessary to fashion a dosage form exhibiting the desired drug release characteristics.

5.10.6   Research and Development

Therics is focusing on commercializing its bone replacement products while continuing to develop drug delivery and tissue engineering technologies. New materials for medical applications are being explored.

5.11 EXTRUDE HONE'S PROMETAL™ 3D PRINTING PROCESS

5.11.1 Company

The ProMetal™ division of Extrude Hone Corporation provides cutting edge rapid prototyping equipment as well as services. It offers users unprecedented flexibility and time savings in creating metal components and tooling for injection molding and casting, utilizing 3D printing technology. With the ability to go directly from CAD data to steel molding inserts in less than two weeks, ProMetal™ offers reduced time to market and greater design latitude for both internal and external geometries. Its address is One Industry Boulevard, Irwin, PA 15642, USA.

5.11.2 Products

5.11.2.1 *Model and Specifications*

The new ProMetal™ R Series machines, R10 and R4 creates complex metal parts and fabricates tools which also come with the optional powder recovery unit that increases the productivity of the ProMetal™ process. The ProMetal™ R10 system is shown in Figure 5.26. Specifications of the machines are summarized in Table 5.9.

5.11.2.2 *Advantages*

(1) *Fast*. The ProMetal™ machine creates multiple parts simultaneously not sequentially like laser systems. Interchangeable build chambers allow quick turnaround between jobs. Build rates can be over 4000 cm^3 per hour.
(2) *Flexible*. Virtually no restriction on design flexibility, such that complex internal geometries and undercuts, can be created.
(3) *Reliable*. There is auto tuning and calibration for maximum performance and also built-in self diagnostics and status reporting. The inkjet printing process is simple and reliable.

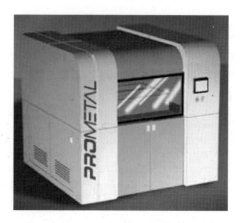

Figure 5.26: The ProMetal™ R10 machine
(Courtesy of Extrude Hone)

Table 5.9: Summarized specifications of ProMetal™ machines

Model	R10 system	R4 system
Build chamber dimensions	1000 mm × 500 mm × 250 mm	400 mm × 400 mm × 250 mm
Printhead options	8, 32 or 96 jet Prometal printhead	8, 16 or 32 jet Prometal printhead
Droplet rate (nozzle/second)	30 000–50 000	30 000–50 000
Droplet placement accuracy	0.025 mm	0.025 mm
Computer system	Pentium III 800 MHz	Pentium III 800 MHz
Software	Windows NT — Prometal software	Windows NT — Prometal software
Power supply	240 V$_{AC}$, 50 A, 50/60 Hz, 3 phase	240 V$_{AC}$, 50 A, 50/60 Hz, 3 phase

(4) *Large parts.* Large steel mold parts measuring 1016 mm × 508 mm × 254 mm can be rapid prototyped.

5.11.3 Process

The ProMetal™ process for three-dimensional printing (3DP) was developed by Extrude Hone Corporation working with experts at MIT. Utilizing 3DP technology, the ProMetal™ process has the ability to build metal components by selectively binding metal powder layer by layer. The finished structural skeleton is then sintered and infiltrated with bronze to produce a finished part that is 60% steel and 40% bronze. The process consists of the following steps. Figure 5.27 illustrates the ProMetal™ process schematically [34].

(1) A part is first designed on a computer using a commercial computer-aided design (CAD) software.
(2) The CAD image is then transferred to the control unit. A very smooth layer of steel powder is then collected from the metal powder supply and spread onto the part build piston.
(3) The CAD image is printed with an ink jet print head depositing millions of droplets of binder per second. These droplets dry quickly upon deposition.
(4) The part build piston lowers approximately 120–170 μm. The process is repeated until the part is completely printed.
(5) The resulting "green" part, of about 60% density, is removed from the machine and excess powder is brushed away.
(6) The "green" part is next sintered in a furnace, while burning off the binder. It is then infiltrated with molten bronze via capillary action to obtain a full density. This is carried out in an infiltration furnace.
(7) Post-processing include machining, polishing and coating to enhance wear and chemical resistance, e.g., nickel and chrome plating.

5.11.4 Principle

The working principle of ProMetal's three-dimensional printing (3DP) is shown in Figure 5.27. It uses an electrostatic ink-jet printing head to

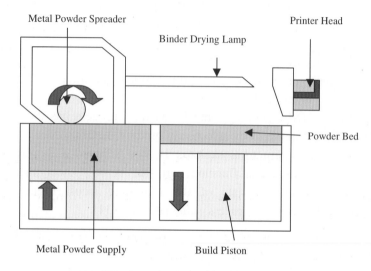

Figure 5.27: The ProMetal™ process

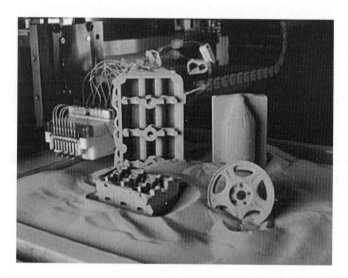

Figure 5.28: Large metal parts produced by the ProMetal™ method
(Courtesy of Extrude Hone Corporation)

deposit a liquid binder onto the powder metals. The part is built one layer at a time based on the sliced cross-sectional data. The metal powder layer is spread on the build piston and a sliced layer is printed onto the powder layer by the ink jet print head depositing droplets of binder that are in turn dried by the binder drying lamp [35]. The process is repeated until the part build is completed.

5.11.5 Applications

The ProMetal™ three-dimensional printing is primarily used to rapidly fabricate complex stainless or tool steel tooling parts. Applications include injection molds, extrusion dies, direct metal components and blow molding [36]. The technology is also suitable for repairing worn out metal tools. Figure 5.28 shows a photograph of large metal parts produced using the technique.

REFERENCES

[1] 3D Systems (DTM Corp.), *Horizons* **Q4** (1999): 6–7.

[2] 3D Systems (DTM Corp.) Product Brochures (material specifications), 1998 to 2001.

[3] Johnson, J.L., *Principles of Computer Automated Fabrication*, Palatino Press, Chapter 3: 75, 1994.

[4] Sun, M.S.M., Nelson, J.C., Beaman, J.J., and Barlow, J.J., "A model for partial viscous sintering," *Proceedings of the Solid Freeform Fabrication Symposium*, University of Texas, 1991.

[5] Hug, W.F. and Jacobs, P.F., "Laser technology assessment for strereolithographic systems," *Proceedings of the Second International Conference on Rapid Prototyping*, June 23–26, 1991: 29–38.

[6] Barlow, J.J., Sun, M.S.M., and Beaman, J.J., "Analysis of selective laser sintering," *Proceedings of the Second International Conference on Rapid Prototyping*, June 23–26, 1991: 29–38.

[7] Spielman, R., "Laser sintered parts in space," *RP = Rapid Production: An International Conference*, held in conjunction with *EuroMold 2000*, Frankfurt, Germany, December 1, 2000.

[8] Anonymous, "Rapid prototyping helps Reebok stay on course," *Rapid Prototyping Report*, November 2000.

[9] Annonymous, "Application of rapid prototyping within Rover Group," *IEE Colloquium (Digest)*, (077), Manufacturing Division Colloquium on Rapid Phototyping in the UK. London, UK, March 23, 1994: 3/1–3/6.

[10] Grochowski, A., "Rapid prototyping — Rapid tooling," *CADCAM Forum*, 39–41, 2000 (http://itri.loyola.edu/rp/02_03.htm).

[11] Serbin, J.C.W., Pretsch, C., and Shellabear, M., *STEREOS and EOSINT 1995 — New Developments and State of the Art*, EOS GmbH, 1995.

[12] ManufacturingTalk.com, "Rapid prototyping goes into production too," October 11, 2001 (http://www.manufacturingtalk.com/news/eos/eos104.htmlFKM Gas Tank EOS).

[13] Solid Caddgroup Inc., Imakenews.com, *Autoinformer* (11), February 6, 2002.

[14] Z Corporation (http://www.zcorp.com).

[15] Z Corporation (http://www.zcorp.com/content/case_studies/shoe.html).

[16] Z Corporation (http://www.zcorp.com/content/case_studies/javelin.html).

[17] NDS.com, "Rapid prototyping to market," *Manufacturing Industry News*, August 27, 2001.

[18] Hofmeister, W., Wert, M., Smugeresky, J., Philliber, J.A., Griffith, M., and Ensz, M., "Investigating Solidification with the Laser-Engineered Net-Shaping (LENS™) Process," *Journal of Materials* **5**(7) (1999).

[19] Uziel, Y., "Art to part in 10 days," *Machine Design*, August 10 (1995): 56–60.

[20] Gregor, A., "From PC to factory," *Los Angeles Times*, October 12, 1994.

[21] Soligen News Release, Soligen's Parts Now™ dazzles Caterpillar with a new dimension in metal parts manufacturing, Soligen Inc., 1995.

[22] Fraunhofer-Gesellschaft, Profile of the Fraunhofer-Gesellschaft: Its purpose, capabilities and prospects, Fraunhofer-Gesellschaft, 1995.

[23] Geiger, M., Steger, W., Greul, M., and Sindel, M., "Multiphase jet solidification, European action on rapid prototyping," *EARP-Newsletter 3*, Aarhus, 1994.

[24] Rapid Prototyping Report, Multiphase Jet Solidification (MJS), 4(6), CAD/CAM Publishing Inc., June, 1994: 5.

[25] Greulick, M., Greul, M., and Pitat, T., "Fast, functional prototypes via Multiphase Jet Solidification," *Rapid Prototyping Journal* **1**(1) (1995): 20–25.

[26] Arcam AB., CAD to Metal® (http://www.arcam.com).

[27] Arcella, F.G., Whitney, E.J., and Krantz, D., "Laser forming near shapes in titanium," *ICALEO '95: Laser Materials Processing*, 80, 1995: 178–183.

[28] Abott, D.H. and Arcella, F.G., "AeroMet implementating novel Ti process," *Metal Powder Report* **53**(2) (February 1998): 24.

[29] Arcella, F.G., Abbot, D.H., and House, M.A., "Rapid laser forming of titanium structures," *Metallurgy World Conference and Exposition*, Grenada, Spain, October 18–22, 1998.

[30] Uziel, Y., Generis Announcement, 5 December 2000 (http://soligen.com/msgbase/messages/741.html).

[31] Therics Inc., Website (http://www.therics.com/theriform.html).

[32] Livingston, T., "Scaffolds for stem-cell based tissue regeneration," *Biomaterials News*, Spring 2002.

[33]   Sharke, P., "Feature focus: Rapid transit to manufacturing," *Mechanical Engineering*, ASME, March 2001: 1–6.

[34]   The Editors, "Added options for producing "impossible" shapes, rapid traverse — Technology and trends spotted by the Editors of Modern Machine Shop," *MMS Online*, 11 September 2001.

[35]   Vasilash G.S., "Feature article: A quick look at rapid prototyping," *Automotive Design and Production*, September 2001: 3.

[36]   Waterman, P.J., "RP3: Rapid prototyping, pattern making, and production," *DE Online*, March 2000.

PROBLEMS

1. Using a sketch to illustrate your answer, describe the Selective Laser Sintering (SLS®) process.

2. Review the types of materials available for the Vanguard™ si2™ SLS® system. What are their respective applications?

3. Describe the principles relating to the SLS® process.

4. Discuss the advantages and disadvantages of 3D System's Vanguard™ si2™ SLS® system.

5. Describe the process flow of Soligen's Direct Shell Production Casting (DSPC).

6. Compare and contrast the laser-based SLS process and the DSPC systems. What are the advantages and disadvantages for each of the systems?

7. What are the critical factors that influence the performance and functionalities of the following RP processes?
   (a) 3D System's SLS,
   (b) Z Corporation's 3DP,
   (c) Soligen's DSPC,
   (d) Fraunhofer's MJS.

8. What are the major differences amongst the various EOS EOSINT models?

9. Compare and contrast EOS's EOSINT machines with its own STEREOS machines.

10. Name three powder-based RP systems that make use of MIT's 3D printing licenses.

11. Compare and contrast EOS's EOSINT M system with Optomec's LENS system. What are the advantages and disadvatages for each of these systems?

12. Discuss the advantages and disadvantages of powder-based RP systems compared with:
    (a) Liquid-based RP systems,
    (b) Solid-based RP systems.

13. Explain how Z Corporation's 3D color printer manufactures multi-colored parts. How do colorized prototypes add value to the RP part?

14. Which powder-based RP machine can accommodate the largest build?

15. Describe Acram's Electron Beam Melting (EBM) technology.

16. Compare and contrast Generis' RP and Extrude Hone's ProMetal™ systems.

17. Explain how Therics' Theriform technology is used for producing parts for tissue engineering and drug delivery parts.

Chapter 6
RAPID PROTOTYPING DATA FORM

6.1 STL FORMAT

Representation methods used to describe CAD geometry vary from one system to another. A standard interface is needed to convey geometric descriptions from various CAD packages to rapid prototyping systems. The STL (**ST**ereoLithography) file, as the *de facto* standard, has been used in many, if not all, rapid prototyping systems.

The STL file [1–3], conceived by the 3D Systems, USA, is created from the CAD database via an interface on the CAD system. This file consists of an unordered list of triangular facets representing the outside skin of an object. There are two formats to the STL file. One is the ASCII format and the other is the binary format. The size of the ASCII STL file is larger than that of the binary format but is human readable. In a STL file, triangular facets are described by a set of X, Y and Z coordinates for each of the three vertices and a unit normal vector with X, Y and Z to indicate which side of facet is an object. An example is shown in Figure 6.1.

Because the STL file is a facet model derived from precise CAD models, it is, therefore, an approximate model of a part. Besides, many commercial CAD models are not robust enough to generate the facet model (STL file) and frequently have problems.

Nevertheless, there are several advantages of the STL file. First, it provides a simple method of representing 3D CAD data. Second, it is already a *de facto* standard and has been used by most CAD systems and rapid prototyping systems. Finally, it can provide small and accurate files for data transfer for certain shapes.

On the other hand, several disadvantages of the STL file exist. First, the STL file is many times larger than the original CAD data file for a given accuracy parameter. The STL file carries much redundancy

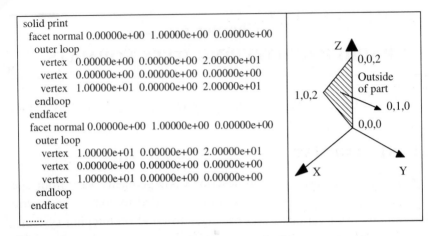

Figure 6.1: A sample STL file

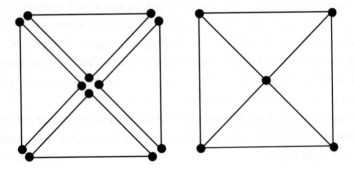

Figure 6.2: Edge and vertex redundancy in STL format

information such as duplicate vertices and edges shown in Figure 6.2. Second, the geometry flaws exist in the STL file because many commercial tessellation algorithms used by CAD vendor today are not robust. This gives rise to the need for a "repair software" which slows the production cycle time. Finally, the subsequent slicing of large STL files can take many hours. However, some RP processes can slice while they are building the previous layer and this will alleviate this disadvantage.

6.2 STL FILE PROBLEMS

Several problems plague STL files and they are due to the very nature of STL files as they contain no topological data. Many commercial tessellation algorithms used by CAD vendors today are also not robust [4–6], and as a result they tend to create polygonal approximation models which exhibit the following types of errors:

(1) Gaps (cracks, holes, punctures) that is, missing facets.
(2) Degenerate facets (where all its edges are collinear).
(3) Overlapping facets.
(4) Non-manifold topology conditions.

The underlying problem is due, in part, to the difficulties encountered in tessellating trimmed surfaces, surface intersections and controlling numerical errors. This inability of the commercial tessellation algorithm to generate valid facet model tessellations makes it necessary to perform model validity checks before the tessellated model is sent to the Rapid Prototyping equipment for manufacturing. If the tessellated model is invalid, procedures become necessary to determine the specific problems, whether they are due to gaps, degenerate facets or overlapping facets, etc.

Early research has shown that repairing invalid models is difficult and not at all obvious [7]. However, before proceeding any further into discussing the procedures that are generated to resolve these difficulties, the following sections shall clarify the problems, as mentioned earlier. In addition, an illustration would be presented to show the consequences brought about by a model having a missing facet, that is, a gap in the tessellated model.

6.2.1 Missing Facets or Gaps

Tessellation of surfaces with large curvature can result in errors at the intersections between such surfaces, leaving gaps or holes along edges of the part model [8]. A surface intersection anomaly which results in a gap is shown in Figure 6.3.

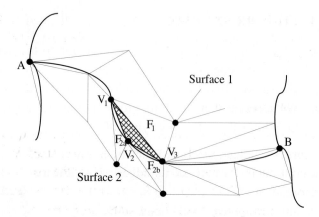

Figure 6.3: Gaps due to missing facets [4]

6.2.2   Degenerate Facets

A geometrical degeneracy of a facet occurs when all of the facets' edges are collinear even though all its vertices are distinct. This might be caused by stitching algorithms that attempt to avoid shell punctures as shown in Figure 6.4(a) below [9].

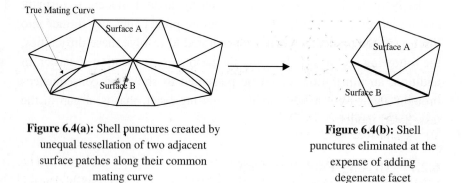

Figure 6.4(a): Shell punctures created by unequal tessellation of two adjacent surface patches along their common mating curve

Figure 6.4(b): Shell punctures eliminated at the expense of adding degenerate facet

The resulting facets generated, shown in Figure 6.4(b), eliminate the shell punctures. However, this is done at the expense of adding a degenerate facet. While degenerate facets do not contain valid surface normals, they do represent implicit topological information on how two surfaces mated. This important information is consequently stored prior to discarding the degenerate facet.

6.2.3 Overlapping Facets

Overlapping facets may be generated due to numerical round-off errors occurring during tessellation. The vertices are represented in 3D space as floating point numbers instead of integers. Thus the numerical round-off can cause facets to overlap if tolerances are set too liberally. An example of an overlapping facet is illustrated in Figure 6.5.

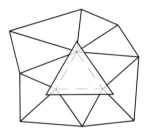

Figure 6.5: Overlapping facets

6.2.4 Non-manifold Conditions

There are three types of non-manifold conditions, namely:

(1) A non-manifold edge.
(2) A non-manifold point.
(3) A non-manifold face.

These may be generated because tessellation of the fine features are susceptible to round-off errors. An illustration of a non-manifold edge is shown in Figure 6.6(a). Here, the non-manifold edge is actually shared by four different facets as shown in Figure 6.6(b). A valid model would be one whose facets have only an adjacent facet each, that is, one edge is shared by two facets only. Hence the non-manifold edges must be resolved such that each facet has only one neighboring facet

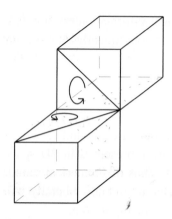

Figure 6.6(a): A non-manifold edge whereby two imaginary minute cubes share a common edge

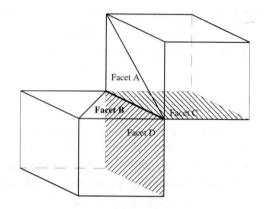

Figure 6.6(b): A non-manifold edge whereby four facets share a common edge after tessellation

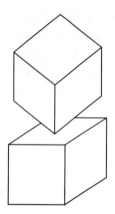

Figure 6.6(c): Non-manifold point

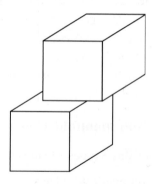

Figure 6.6(d): Non-manifold face

along each edge, that is, by reconstructing a topologically manifold surface [4]. In Figures 6.6(c) and 6.6(d), two other types of non-manifold conditions are shown.

All problems that have been mentioned previously are difficult for most slicing algorithms to handle and they do cause fabrication problems for RP processes which essentially require valid tessellated solids as input. Moreover, these problems arise because tessellation is a first-order approximation of more complex geometric entities. Thus, such problems have become almost inevitable as long as the representation of the solid model is done using the STL format which inherently has these limitations.

6.3 CONSEQUENCES OF BUILDING A VALID AND INVALID TESSELLATED MODEL

The following sections present an example each of the outcome of a model built using a valid and an invalid tessellated model as an input to the RP systems.

6.3.1 A Valid Model

A tessellated model is said to be valid if there are no missing facets, degenerate facets, overlapping facets or any other abnormalities. When a valid tessellated model (see Figure 6.7(a)) is used as an input, it will first be sliced into 2D layers, as shown in Figure 6.7(b). Each layer would then be converted into unidirectional (or 1D) scan lines for the laser or other RP techniques to commence building the model as shown in Figure 6.7(c).

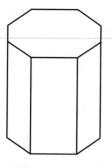
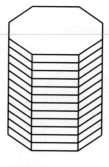
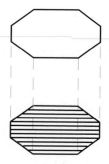

Figure 6.7(a): A valid 3D model

Figure 6.7(b): A 3D model sliced into 2D planar layers

Figure 6.7(c): Conversion of 2D layers into 1D scan lines

The scan lines would act as on/off points for the laser beam controller so that the part model can be built accordingly without any problems.

6.3.2   An Invalid Model

However, if the tessellated model is invalid, a situation may develop as shown in Figure 6.8.

A solid model is tessellated non-robustly and results in a gap as shown in Figure 6.8(a). If this error is not corrected and the model is subsequently sliced, as shown in Figure 6.8(b), in preparation for it to

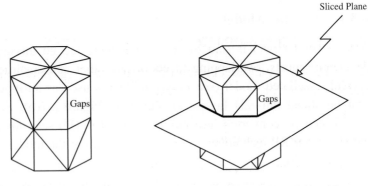

Figure 6.8(a): An invalid tessellated model

Figure 6.8(b): An invalid model being sliced

TOP VIEW

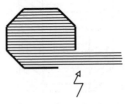

Stray Scan-Vectors

Figure 6.8(c): A layer of an invalid model being scanned

be built layer by layer, the missing facet in the geometrical model would cause the system to have no predefined stopping boundary on the particular slice, thus the building process would continue right to the physical limit of the RP machine, creating a stray physical solid line and ruining the part being produced, as illustrated in Figure 6.8(c).

Therefore, it is of paramount importance that the model be "repaired" before it is sent for building. Thus, the model validation and repair problem is stated as follows:

> *Given a facet model (a set of triangles defined by their vertices), in which there are gaps, i.e., missing one or more sets of polygons, generate "suitable" triangular surfaces which "fill" the gaps [4].*

6.4 STL FILE REPAIR

The STL file repair can be implemented using a generic solution and dedicated solutions for special cases.

6.4.1 Generic Solution

In order to ensure that the model is valid and can be robustly tessellated, one solution is to check the validity of all the tessellated triangles in the model. This section presents the basic problem of missing facets and a proposed generic solution to solve the problem with this approach.

In existing RP systems, when a punctured shell is encountered, the course of action taken usually requires a skilled technician to manually repair the shell. This manual shell repair is frequently done without any knowledge of the designer's intent. The work can be very time-consuming and tedious, thus negating the advantages of rapid prototyping as the cost would increase and the time taken might be longer than that taken if traditional prototyping processes were used.

The main problem of repairing the invalid tessellated model would be that of matching the solution to the designer's intent when it may have been lost in the overall process. Without the knowledge of the designer's intent, it would indeed be difficult to determine what the "right" solution should be. Hence, an "educated" guess is usually made when faced with ambiguities of the invalid model. The algorithm in this report aims to match, if not exceed, the quality of repair done manually by a skilled technician when information of the designer's intent is not available.

The basic approach of the algorithm to solve the "missing facets" problem would be to detect and identify the boundaries of all the gaps in the model. Once the boundaries of the gap are identified, suitable facets would then be generated to repair and "patch up" these gaps. The size of the generated facets would be restricted by the gap's boundaries while the orientation of its normal would be controlled by comparing it with the rest of the shell. This is to ensure that the generated facets' orientation are correct and consistent throughout the gap closure process.

The orientation of the shell's facets can be obtained from the STL file which lists its vertices in an ordered manner following Mobius' rule. The algorithm exploits this feature so that the repair carried out on the invalid model, using suitably created facets, would have the correct orientation.

Thus, this generic algorithm can be said to have the ability to make an inference from the information contained in the STL file so that the following two conditions can be ensured:

(1) The orientation of the generated facet is correct and compatible with the rest of the model.
(2) Any contoured surface of the model would be followed closely by the generated facets due to the smaller facet generated. This is in contrast to manual repair whereby, in order to save time, fewer facets generated to close the gaps are desired, resulting in large generated facets that do not follow closely to the contoured surfaces.

Finally, the basis for the working of the algorithm is due to the fact that in a valid tessellated model, there must only be two facets sharing every edge. If this condition is not fulfilled, then this indicates that there are some missing facets. With the detection and subsequent repair of these missing facets, the problems associated with the invalid model can then be eliminated.

6.4.1.1 Solving the "Missing Facets" Problem

The following procedure illustrates the detection of gaps in the tessellated model and its subsequent repair. It is carried out in four steps.

(1) Step 1: Checking for Approved Edges with Adjacent Facets

The checking routine executes as follows for Facet A as seen in Figure 6.9:

(a)   (i) Read in first edge {vertex 1-2} from the STL file.
 (ii) Search file for a similar edge in the opposite direction {vertex 2-1}.
 (iii) If edge exists, store this under a temporary file (e.g., file B) for approved edges.
 (iv) Do the same for 2 and 3 below.

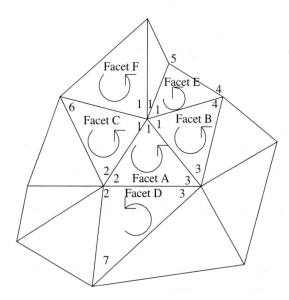

Figure 6.9: A representation of a portion of a tessellated surface without any gaps

(b)　(i)　Read in second edge {vertex 2-3} from the STL file.

　　(ii)　Search file for a similar edge in the opposite direction {vertex 3-2}.

　　(iii)　Perform as in (a)(iii) above.

(c)　(i)　Read in third {vertex 3-1} from the STL file.

　　(ii)　Search file for a similar edge in the opposite direction {vertex 1-3}.

　　(iii)　Perform as in (a)(iii) above.

This process is repeated for the next facet until all the facets have been searched.

(2)　Step 2: Detection of Gaps in the Tessellated Model

The detection routine executes as follows:

For Facet A (please refer to Figure 6.10):

(a)　(i)　Read in edge {vertex 2-3} from the STL file.

　　(ii)　Search file for a similar edge in the opposite direction {vertex 3-2}.

　　(iii)　If edge does not exist, store edge {vertex 3-2} in another temporary file (e.g., file C) for suspected gap's bounding edges and store vertex 2-3 in file B1 for existing edges without adjacent facets (this would be used later for checking the generated facet orientation).

For Facet B,

(b)　(i)　Read in edge {vertex 5-2} from the STL file.

　　(ii)　Search file for a similar edge in the opposite direction {vertex 2-5}.

　　(iii)　If it does not exist, perform as in (a)(iii) above.

(c)　(i)　Repeat for edges:　5-2 ; 7-5 ; 9-7 ; 11-9 ; 3-11.

　　(ii)　Search for edges:　2-5 ; 5-7 ; 7-9 ; 9-11 ; 11-3.

　　(iii)　Store all the edges in that temporary file B1 for edges without any adjacent facet and store all the suspected bounding edges of the gap in temporary file C. File B1 can appear as in Table 6.1.

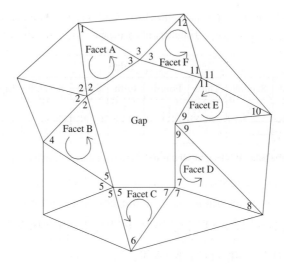

Figure 6.10: A representation of a portion of a tessellated surface with a gap present

Table 6.1: File B1 contains existing edges without adjacent facets

Vertex	Edge					
	First	Second	Third	Fourth	Fifth	Sixth
First	2	7	3	5	9	11
Second	3	5	11	2	7	9

(3) Step 3: Sorting of Erroneous Edges into a Closed Loop

When the checking and storing of edges (both with and without adjacent facets) are completed, a sort would be carried out to group all the edges without adjacent facets to form a closed loop. This closed loop would represent the gap detected and be stored in another temporary file (e.g., file D) for further processing. The following is a simple illustration of what could be stored in file C for edges that do not have an adjacent edge.

Assuming all the "erroneous" edges are stored according to the detection routine (see Figure 6.10 for all the erroneous edges), then file C can appear as in Table 6.2.

Table 6.2: File C containing all the "Erroneous" edges that would form the boundary of each gaps

Vertex	Edge								
	First	Second	Third	Fourth	Fifth	Sixth	Seventh	Eighth	Ninth
First	3	5	*	11	2	7	*	9	*
Second	2	7	*	3	5	9	*	11	*

*Represent all the other edges that would form the boundaries of other gaps

As can be seen in Table 6.2, all the edges are unordered. Hence, a sort would have to be carried out to group all the edges into a closed loop. When the edges have been sorted, it would then be stored in a temporary file, say file D. Table 6.3 is an illustration of what could be stored in file D.

Table 6.3: File D containing sorted edges

	First edge	Second edge	Third edge	Fourth edge	Fifth edge	Sixth edge
First vertex	3	2	5	7	9	11
Second vertex	2	5	7	9	11	3

Figure 6.11 is a representation of the gap, with all the edges forming a sorted closed loop.

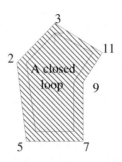

Figure 6.11: A representation of a gap bounded by all the sorted edges

(4) Step 4: Generation of Facets for the Repair of the Gaps

When the closed loop of the gap is established with its vertices known, facets are generated one at a time to fill up the gap. This process is summarized in Table 6.4 and illustrated in Figure 6.12.

Table 6.4: Process of facet generation

		V3	V2	V5	V7	V9	V11
Generation of facets	F1	1	2	—	—	—	3
	F2	E	1	—	—	2	3
	F3	E	1	2	—	3	E
	F4	E	E	1	2	3	E

V = vertex, F = facet, E = eliminated from the process of facet generation

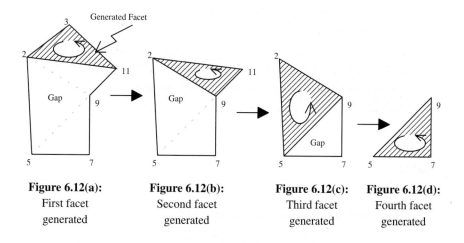

Figure 6.12(a):	**Figure 6.12(b):**	**Figure 6.12(c):**	**Figure 6.12(d):**
First facet generated	Second facet generated	Third facet generated	Fourth facet generated

With reference to File D,

(a) Generating the first facet: First two vertices (V3 and V2) in the first two edges of file D will be connected to the first vertex in the last edge (V11) in file D and the facet is stored in a temporary file E (see Table 6.5 on how the first generated facet would be stored in file E). The facet is then checked for its orientation using the information stored in file B1. Once its orientation is determined to be correct, the first vertex (V3) from file D will be temporarily removed.

(b) Generating the second facet: Of the remaining vertices in file D, the previous second vertex (V2) will become the first edge of file D. The second facet is formed by connecting the first vertex (V2) of the first edge with that of the last two vertices in file D (V9, V11), and the facet is stored in temporary file E. It is then checked to confirm if its orientation is correct. Once it is determined to be correct, the vertex (V11) of the last edge in file D is then removed temporarily.

(c) Generating the third facet: The whole process is repeated as it was done in the generation of facets 1 and 2. The first vertex of the first two edges (V2, V5) is connected to the first vertex of the last edge (V9) and the facet is stored in temporary file E. Once its orientation is confirmed, the first vertex of the first edge (V2) will be removed from file D temporarily.

(d) Generating the fourth facet: The first vertex in the first edge will then be connected to the first vertices of the last two edges to form the fourth facet and it will again be stored in the temporary file E. Once the number of edges in file D is less than three, the process of facet generation will be terminated. After the last facet is generated, the data in file E will be written to file A and its content (file E's) will be subsequently deleted. Table 6.5 shows how file E may appear.

Table 6.5: Illustration of how data could be stored in File E

Generated facet	First edge		Second edge		Third edge	
	First vertex	Second vertex	First vertex	Second vertex	First vertex	Second vertex
First	V3	V2	V2	V5	V5	V3
Second	V2	V9	V9	V11	V11	V2
Third	V2	V5	V5	V9	V9	V2
Fourth	V5	V7	V7	V9	V9	V5

The above procedures work for both types of gaps whose boundaries consist either of odd or even number of edges. Figure 6.13 and Table 6.6 illustrate how the algorithm works for an *even* number of edges or vertices in file D.

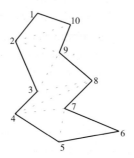

Figure 6.13: Gaps with even number of edges

Table 6.6: Process of facet generation for gaps with even number of edges

Facets	Vertices									
	V1	V2	V3	V4	V5	V6	V7	V8	V9	V10
F1	1	2								3
F2	E	1							2	3
F3	E	1	2						3	E
F4	E	E	1					2	3	E
F5	E	E	1	2				3	E	E
F6	E	E	E	1			2	3	E	E
F7	E	E	E	1	2		3	E	E	E
F8	E	E	E	E	1	2	3	E	E	E

With reference to Table 6.6,

First facet generated:
Edge 1 → V1, V2
Edge 2 → V2, V10
Edge 3 → V10, V1

Second facet generated:
Edge 1 → V2, V9
Edge 2 → V9, V10
Edge 3 → V10, V1

and so on until the whole gap is covered. Similarly, Figure 6.14 and Table 6.7 illustrate how the algorithm works for an *odd* number of edges or vertices in file D.

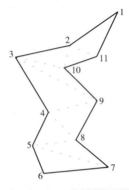

Figure 6.14: Gaps with odd number of edges

Table 6.7: Process of facet generation for gaps with odd number of edges

Facets	\multicolumn{11}{c}{Vertices}

Facets	V1	V2	V3	V4	V5	V6	V7	V8	V9	V10	V11
F1	1	2									3
F2	E	1								2	3
F3	E	1	2							3	E
F4	E	E	1						2	3	E
F5	E	E	1	2					3	E	E
F6	E	E	E	1				2	3	E	E
F7	E	E	E	1	2			3	E	E	E
F8	E	E	E	E	1		2	3	E	E	E
F9	E	E	E	E	1	2	3	E	E	E	E

The process of facet generation for *odd* vertices are also done in the same way as *even* vertices. The process of facet generation has the following pattern:

F1 → First and second vertices are combined with the last vertex. Once completed, eliminate first vertex. The remainder is ten vertices.

F2 → First vertex is combined with the last two vertices. Once completed, eliminate the last vertex. The remainder is nine vertices.

F3 → First and second vertices are combined with the last vertex. Once completed, eliminate first vertex. The remainder is eight vertices.

F4 → First vertex is combined with last two vertices. Once completed, eliminate the last vertex. The remainder is seven vertices.

This process is continued until all the gaps are patched.

6.4.1.2 *Solving the "Wrong Orientation of Facets" Problem*

In the case when the generated facet's orientation is wrong, the algorithm should be able to detect it and corrective action can be taken to rectify this error. Figure 6.15 shows how a generated facet with a wrong orientation can be corrected.

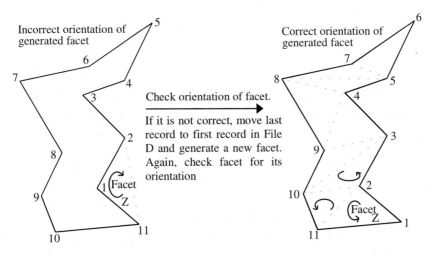

Figure 6.15: Incorrectly generated facet's orientation and its repair

It can be seen that facet Z (vertices 1, 2, 11) is oriented in a clockwise direction and this contradicts the right-hand rule adopted by the STL format. Thus, this is not acceptable and needs corrections.

This can be done by shifting the last record in file D of Table 6.8 to the position of the first edge in file D of Table 6.9. All the edges, including the initial first one will be shifted one position to the right (assuming that the records are stored in the left to right structure). Once this is done, step 4 of facet generation can be implemented.

Before the shift:

Table 6.8: Illustration showing how file D is manipulated to solve orientation problems

Vertex	Edge										
	First	Second	Third	Fourth	Fifth	Sixth	Seventh	Eighth	Ninth	Tenth	Eleventh
First	1	2	3	4	5	6	7	8	9	10	11
Second	2	3	4	5	6	7	8	9	10	11	1

After the shift:

Table 6.9: Illustration showing the result of the shift to correct the facet orientation

Vertex	Edge										
	First	Second	Third	Fourth	Fifth	Sixth	Seventh	Eighth	Ninth	Tenth	Eleventh
First	11	1	2	3	4	5	6	7	8	9	10
Second	1	2	3	4	5	6	7	8	9	10	11

As can be seen from the above example, vertices 1 and 2 are used initially as the first edge to form a facet. However, this resulted in a facet having a clockwise direction. After the shift, vertices 11 and 1 are used as the first edge to form a facet.

Facet Z, as shown on the right-hand-side of Figure 6.15, is again generated (vertices 1, 2, 11) and checked for its orientation. When its orientation is correct (i.e., in the anti-clockwise direction), it is saved and stored in temporary file E.

All subsequent facets are then generated and checked for its orientation. If any of its subsequently generated facets has an incorrect orientation, the whole process would be restarted using the initial temporary file D. If all the facets are in the right orientation, it will then be written to the original file A.

6.4.1.3 *Comparison with an Existing Algorithm that Performs Facet Generation*

An illustration of an existing algorithm that might cause a very narrow facet (shaded) to be generated is shown in Figure 6.16.

This results from using an algorithm that uses the smallest angle to generate a facet. In essence, the problem is caused by the algorithm's search for a time-local rather than a global-optimum solution [10]. Also, calculation of the smallest angle in 3D space is very difficult.

Figure 6.16 is similar to Figure 6.14. However, in this case, the facet generated (shaded) can be very narrow. In comparing the algorithms, the result obtained would match, if not, exceed the algorithm that uses the smallest angle to generate a facet.

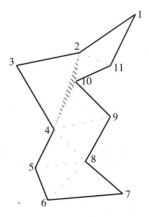

Figure 6.16: Generation of facets using an algorithm that uses the smallest angle between edges

6.4.2   Special Algorithms

The generic solution presented could only cater to gaps (whether simple or complex) that are isolated from one another. However, should any of the gaps were to meet at a common vertex, the algorithm may not be able to work properly. In this section, the algorithm would be expanded to include solving some of these special cases. These special cases include:

(1) Two or more gaps formed from a coincidental vertex.
(2) Degenerate facets.
(3) Overlapping facets.

The special cases are classified as such because these errors are not commonly encountered in the tessellated model. Hence it is not advisable to include this expanded algorithm in the generic solution as it can be very time consuming to apply in a during normal search. However, if there are still problems in the tessellated model after the generic solution's repair, the expanded algorithm could then be used to detect and solve the special case problems.

6.4.2.1   *Special Case 1: Two or More Gaps are Formed from a Coincidental Vertex*

The first special case deals with problems where two or more gaps are formed from a coincidental vertex. Appropriate modifications to the general solution may be made according to the solutions discussed as follows.

As can be seen from Figure 6.17, there exists two gaps that are connected to vertex 1. The algorithm given in Paper I would have a problem identifying which vertex to go to when the search reaches vertex 1 (either vertex 2 in gap 1 or vertex 5 in gap 2).

Table 6.10 illustrates what file C would look like, given the two gaps.

When the search starts to find all the edges that would form a closed loop, the previous algorithm might mistakenly connect edges: 3-4; 4-1; and *1-5*; instead of *1-2*. This is clearly an error as the edge that is

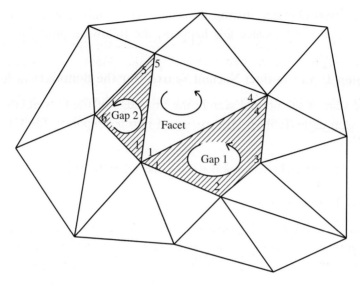

Figure 6.17: Two gaps sharing one coincidental vertex

Table 6.10: File C containing "Erroneous" edges that would form the
boundaries of gaps

File C :

Vertex	**Edge**						
	First	Second	Third	Fourth	Fifth	Sixth	Seventh
First	V3	V5	V4	*V1*	V6	V2	*V1*
Second	V4	V6	*V1*	*V5*	V1	V3	*V2*

supposed to be included in file D should be *1-2, and not 1-5*. It is
therefore pertinent that for every edge searched, the second vertex (e.g.,
V1 of the third edge, shaded, in Table 6.10) of that edge should be
searched against the first vertex of subsequent edges (e.g., V1 of edge
1-5 in the fourth edge) and this should not be halted the first time the
first vertex of subsequent edges are found. The search should continue
to check if there are other edges with V1 (e.g., edge 1-2, in the seventh
edge). Every time, say for example, vertex 1 is found, there should be a
count. When the count is more than two, that would indicate that there

is more than one gap sharing the same vertex; this may be called the coincidental vertex. Once this happens, the following procedure would then be used.

(1) Step 1: To Conduct Normal Search (for the Boundary of Gap 1)

At the start of the normal search, the first edge of file C, vertices 3 and 4, as seen as in Table 6.11(a), is saved into a temporary file C1.

Table 6.11(a): Representation of how files C and C1 would look like

File C:

Vertex	Edge						
	First	Second	Third	Fourth	Fifth	Sixth	Seventh
First	V3	V5	V4	V1	V6	V2	V1
Second	V4	V6	V1	V5	V1	V3	V2

File C1:

Vertex	Edge		
	First	Second	Third
First	V3	?	?
Second	V4	?	?

The second vertex in the first edge (V4) of file C1, is searched against the first vertex of subsequent edge in file C (refer to Table 6.11(a), shaded box). Once it is found to be the same vertex (V4) and that there are no other edges sharing the same vertex, the edge (i.e., vertex 4-1) is stored as the second edge in file C1 (refer to Table 6.11(b)).

(2) Step 2: To Detect More Than One Gap

The second vertex of the second edge (V1) in file C1 is searched for an equivalent first vertex of subsequent edges in file C (refer to Table 6.11(b), shaded box containing vertex). Once it is found (V1, first vertex of fourth edge), a count of one is registered and at the same time, that edge is noted.

Table 6.11(b): Representation on how files C and C1 would look like
during the normal search (special case)

File C:

				Count 1 ⇓			Count 2 ⇓
	Edge						
Vertex	First	Second	Third	Fourth	Fifth	Sixth	Seventh
First	V3	V5	V4	V1	V6	V2	V1
Second	V4	V6	V1	V5	V1	V3	V2

File C1:

	Edge		
Vertex	First	Second	Third
First	V3	V4	?
Second	V4	V1	?

The search for the same vertex is continued to determine if there are other edges sharing the same vertex 1. If there is an additional edge sharing the same vertex 1 (Table 6.11(b), seventh edge), another count is registered, making a total of two counts. Similarly, the particular record in which it happens again is noted. The search is continued until there are no further edges sharing the same vertex. When completed, the following is carried out for the third edge in file C1.

For the first count, reading from file C, all the edges that would form the first alternative closed loop are sorted. These first alternatives in file C2 (refer to Table 6.11(c) and Figure 6.18(a) for a graphical representation of how the first alternatives might look like) are then saved.

A closed loop is established once the second vertex of the last edge (V3) is the same as the first vertex of the first edge (V3). *For the second count*, the edges are sorted to form a second alternative of a closed loop that will represent the boundary of the gap. These edges are then saved in another temporary file C3 (refer to Table 6.11(d) and Figure 6.18(b) for a graphical representation of how the second

Table 6.11(c): First alternative closed loop that may represent the boundary of the gap

File C2:

Vertex	Edge						
	First	Second	Third	Fourth	Fifth	Sixth	Seventh
First	V3	V4	V1	V5	V6	V1	V2
Second	V4	V1	V5	V6	V1	V2	V3

Table 6.11(d): Second alternative closed loop that may represent the boundary of the gap

File C3:

Vertex	Edge			
	First	Second	Third	Fourth
First	V3	V4	V1	V2
Second	V4	V1	V2	V3

Arrows indicating the direction of loop forming the boundary of gap

Arrows indicating the direction of loop forming the boundary of gap

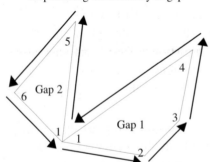

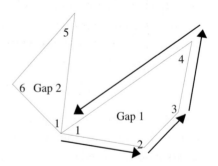

Figure 6.18(a): Graphical representation of the two gaps sharing a coincidental vertex (first alternative)

Figure 6.18(b): Graphical representation of the two gaps sharing the same vertex (second alternative)

alternatives may look like). Once a closed loop is established, the search is stopped.

(3) Step 3: To Compare Which Alternative Has the Least Record to Form the Boundary of the Gap

From file C2 (first alternative) and file C3 (second alternative), it can be seen that the second alternative has the least record to form the boundary of Gap 1. Hence the second alternative data would be written to file D for the next stage of facet generation and the two temporary files C2 and C3 would be discarded.

Once gap 1 is repaired, gap 2 can be repaired by using the generic solution.

6.4.2.2 *Further Ilustrations on How the Algorithm Works*

The algorithm can cater to more than two gaps sharing the same vertex, as shown in Figure 6.19(a), or even three gaps arranged differently, as shown in Figure 6.19(b).

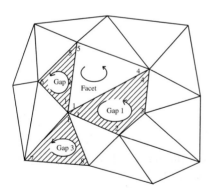
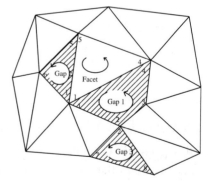

Figure 6.19(a): Three facets sharing one coincidental vertex

Figure 6.19(b): Three facets sharing two coincidental vertices

The solution is illustrated using the condition shown in Figure 6.20(b). Table 6.12(a) shows how file C may appear.

Table 6.12(a): Illustration on how file C may appear given three gaps sharing the same two vertices

File C:

Vertex	Edge									
	First	Second	Third	Fourth (Count 1 for V1)	Fifth	Sixth (Count 1 for V2)	Seventh (Count 2 for V1)	Eighth (Count 2 for V2)	Ninth	Tenth
First	V3	V5	V4	V1	V6	V2	V1	V2	V8	V7
Second	V4	V6	V1	V5	V1	V3	V2	V7	V2	V8

(1) Step 1: Normal Search

Table 6.12(b): File C1 during normal search

File C1:

Vertex	Edge			
	First	Second	Third	Fourth
First	V3	V4	?	?
Second	V4	V1	?	?

(2) Step 2: Detection

Referring to Table 6.12(a), it can be seen that:

For V1 → there are two counts
For V2 → there are two counts

<u>For Count 1 for V1 and Count 1 for V2</u>

Thus from file C (Table 6.12(a)) the first alternative closed loop can be generated and the file is shown in Table 6.12(c).

Table 6.12(c): First alternative of closed loop that may represent the boundary of a gap

Vertex	Edge									
	First	Second	Third	Fourth	Fifth	Sixth	Seventh	Eighth	Ninth	Tenth
First	V3	V4	V1	V5	V6	V1	V2	—	—	—
Second	V4	V1	V5	V6	V1	V2	V3	—	—	—

Figure 6.20(a) illustrates a graphical representation of the gap's boundary.

For Count 1 for V1 and Count 2 for V2

The second alternative of a closed loop can be sorted and is shown in Table 6.12(d).

Table 6.12(d): Second alternative of closed loop that may represent the boundary of a gap

Vertex	Edge									
	First	Second	Third	Fourth	Fifth	Sixth	Seventh	Eighth	Ninth	Tenth
First	V3	V4	V1	V5	V6	V1	V2	V7	V8	V2
Second	V4	V1	V5	V6	V1	V2	V7	V8	V2	V3

Figure 6.20(b) illustrates a graphical representation of the gap's boundary.

For Count 2 for V1 and Count 1 for V2

The third alternative of a closed loop can again be sorted and is shown in Table 6.12(e).

Table 6.12(e): Third alternative of closed loop that may represent the boundary of a gap

Vertex	Edge									
	First	Second	Third	Fourth	Fifth	Sixth	Seventh	Eighth	Ninth	Tenth
First	V3	V4	V1	V2	—	—	—	—	—	—
Second	V4	V1	V2	V3	—	—	—	—	—	—

Figure 6.20(c) illustrates a graphical representation of the gap's boundary.

For Count 2 for V1 and Count 2 for V2

The fourth alternative of a closed loop can also be sorted and is shown in Table 6.12(f).

Table 6.12(f): Fourth alternative of closed loop that may represent the boundary of a gap

	Edge									
Vertex	First	Second	Third	Fourth	Fifth	Sixth	Seventh	Eighth	Ninth	Tenth
First	V3	V4	V1	V2	V7	V8	V2	—	—	—
Second	V4	V1	V2	V7	V8	V2	V3	—	—	—

Figure 6.20(d) illustrates a graphical representation of the gap's boundary.

(3) Step 3: Comparison of the Four Alternatives

As can be seen from the four alternatives (see Figure 6.20), the third alternative, as shown in Figure 6.20(c), is considered the best solution and the correct solution to fill gap 1. This correct solution can be found by comparing which alternative uses the least edges to fill gap 1 up. Once the solution is found, the edges would be saved to file D for the next stage, that of facet generation.

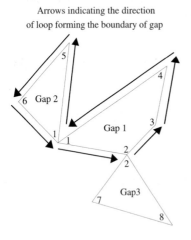

Figure 6.20(a): First alternative —
graphical representation of the three
gaps sharing two coincident vertices

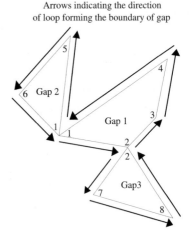

Figure 6.20(b): Second alternative —
graphical representation of the three
gaps sharing two coincident vertices

Arrows indicating the direction
of loop forming the boundary of gap

Arrows indicating the direction
of loop forming the boundary of gap

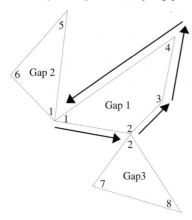

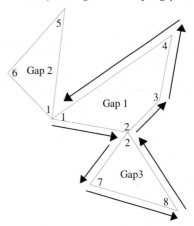

Figure 6.20(c): Third alternative —
graphical representation of the three
gaps sharing two coincident vertices

Figure 6.20(d): Fourth alternative —
graphical representation of the three
gaps sharing two coincident vertices

After gap 1 is filled, gaps 2 and 3 can then be repaired using the basic generic solution.

6.4.2.3 *Special Case 2: Two Facets Sharing a Common Edge*

When a degenerate facet such as the one shown in Figure 6.21 is encountered, vector algebra is applied. The following steps are taken:

(1) Edge a-c is converted into vectors,

$$(c_1\mathbf{i} + c_2\mathbf{j} + c_3\mathbf{k}) - (a_1\mathbf{i} + a_2\mathbf{j} + a_3\mathbf{k}) \qquad (6.1)$$

(2) (i) Collinear vectors are checked.
  Let

$$\mathbf{x} = ac \Rightarrow x_1\mathbf{i} + x_2\mathbf{j} + x_3\mathbf{k}; \qquad (6.2)$$

$$\mathbf{y} = ce \Rightarrow y_1\mathbf{i} + y_2\mathbf{j} + y_3\mathbf{k}; \qquad (6.3)$$

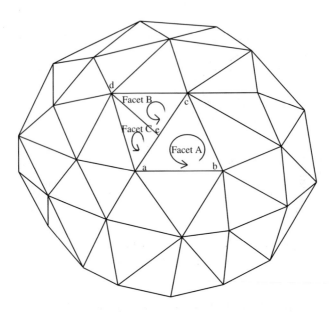

Figure 6.21: An illustration showing a degenerate facet

(ii) By mathematical definition, the two vectors **x** and **y** are said to be collinear if there exists scalars s and t, both non-zero, such that:

$$s\mathbf{x} + t\mathbf{y} = 0. \tag{6.4}$$

But when applied in a computer, it is only necessary to have:

$$s\mathbf{x} + t\mathbf{y} \leq \varepsilon, \tag{6.5}$$

where ε is a definable tolerance.

(3) If the two vectors are found to be collinear vectors, the position of vertex e is generated.

(4)　(i) Facet A is split into two facets (see Figure 6.22). The two facets are generated using the three vertices in facet A.
First facet vertices are: a, b, e
Second facet vertices are: b, c, e

(ii) The orientation is checked.

(iii) New facets are stored in a temporary file.

(5) Search and delete facet A from original file.

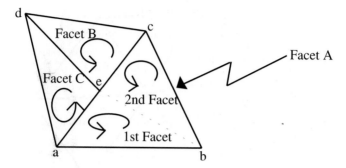

Figure 6.22: Illustration on how a degenerate facet is solved

(6)   (i) The two new facets are stored in the original file.
      (ii) Data are deleted from the temporary file.

6.4.2.4   *Special Case 3: Overlapping Facets*

The condition of overlapping facets can be caused by errors introduced by inconsistent numerical round-off. This problem can be resolved through vertex merging where vertices within a predetermined numerical round-off tolerance of one another can be merged into just one vertex. Figure 6.23 illustrates one example of how this solution can be applied. Figure 6.24 illustrates another example of an overlapping facet.

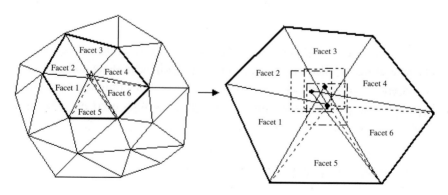

Figure 6.23(a): Overlapping facets

Figure 6.23(b): Numerical round-off equivalence region

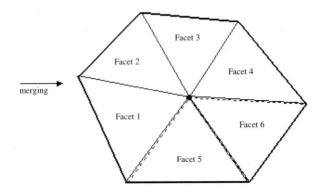

Figure 6.23(c): Vertices merged

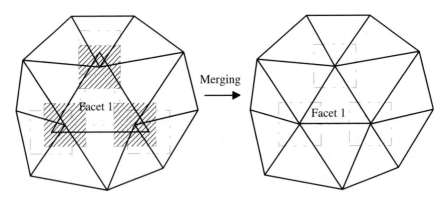

Figure 6.24(a): An overlapping facet

Figure 6.24(b): Facet's vertices merged
with vertices of neighboring facets

It is recommended that this merging of vertices be done before the searching of the model for gaps. This will eliminate unnecessary detection of erroneous edges and save substantial computational time expended in checking whether the edges can be used to generate another facet.

6.4.3   Performance Evaluation

Computational efficiency is an issue whenever CAD-model repair of solids that have been finely tessellated is considered. This is due to the fact that for every unit increase in the number of facets (finer tessellation), the additional increase in the number of edges is three. Thus, the computational time required for the checking of erroneous edges would correspondingly increase.

6.4.3.1   *Efficiency of the Detection Routine*

Assuming that there are twelve triangles in the cube (Figure 6.25), the number of edges = $12 \times 3 = 36$. The number of searches is computed as follows:

(1) Read first edge, search 35 and remove two edges.
(2) Read second edge, search 33 and remove two edges.
(3) Read third edge, search 31 and remove two edges.
(4) Number of searches = $1 + 3 + \cdots + 35$
    In general, the number of searches = $1 + \cdots + (n - 1)$
$$= n^2/4 \qquad\qquad (6.6)$$
    where n = number of edges.

Although this result does not seem satisfactory, it is both an optimum and a robust solution that can be obtained given the inherent nature of the STL format (such as its lack of topological information).

However, if topological information is available, the efficiency of the routine used to detect the erroneous edges can definitely be increased significantly. Some additional points worth noting are: First, binary files are far more efficient than ASCII files because they are only 20–25% as large and thus reduce the amount of physical data that needs to be transferred and because they do not require subsequent

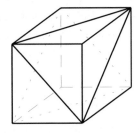

Figure 6.25: A cube tessellated into 12 triangles

translation into a binary representation. Consequently, one easily saves several minutes per file by using binary instead of ASCII file formats [10].

As for vertex merging, the use of one-dimensional AVL-tree can significantly reduce the search-time for sufficiently identical vertices [10]. The AVL-trees, which are usually twelve to sixteen levels deep, reduces each search from $0(n)$ to $0(\log n)$ complexity, and the total search-time from close to an hour to less than a second.

6.4.3.2   *Estimated Computational Time for Shell Closure*

The computational time required for shell closure is relatively fast. Estimated time can range from a few seconds to less than a minute and is arrived at based on the processing time obtained by Jan Helge Bohn [10] that uses a similar shell closure algorithm.

6.4.3.3   *Limitations of Current Shell Closure Process*

The shell closure process developed thus far does not have the ability to detect or solve the problems posed by any of the non-manifold conditions. However, the detection of non-manifold conditions and their subsequent solutions would be the next focus of the ongoing research.

A limitation of the algorithm involves the solving of co-planar (see Figure 6.26(a)) and non co-planar facets (see Figure 6.26(b)) whose intersections result in another facet. The reasons for such errors are related to the application that generated the faceted model, the application that generated the original 3D CAD model, and the user.

Another limitation involves the incorrect triangulation of parametric surface (see Figure 6.27). One of the overlapping triangles, $T_b = BCD$ should not be present and should thus be removed while the other triangle, $T_a = ABC$ should be split into two triangles so as to maintain the correct contoured surface. The proposed algorithm is presently unable to solve this problem [11].

Finally, as mentioned earlier, the efficiency of $n^2/4$ (when n is the number of edges) is a major limitation especially when the number of

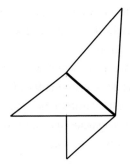 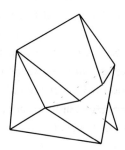

Figure 6.26(a): Incorrect triangulation (co-planar facet)   **Figure 6.26(b):** Non co-planar whereby facets are split after being intersected

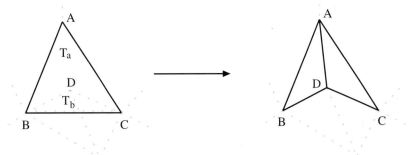

Figure 6.27: Incorrect triangulation of parametric surface

facets in the tessellated model becomes very large (e.g., greater than 40 000). There are differences in the optimum use of computational resources. However, further work is being carried out to ease this problem by using available topological information which are available in the original CAD model.

6.5   OTHER TRANSLATORS

6.5.1   IGES File

IGES (Initial Graphics Exchange Specification) is a standard used to exchange graphics information between commercial CAD systems. It was set up as an American National Standard in 1981 [12, 13]. The IGES file can precisely represent CAD models. It includes not only the geometry information (Parameter Data Section) but also topological information (Directory Entry Section). In the IGES, surface modeling, constructive solid geometry (CSG) and boundary representation (B-rep) are introduced. Especially, the ways of representing the regularized operations for union, intersection, and difference have also been defined.

The advantages of the IGES standard are its wide adoption and comprehensive coverage. Since IGES was set up as American National Standard, virtually every commercial CAD/CAM system has adopted IGES implementations. Furthermore, it provides the entities of points, lines, arcs, splines, NURBS surfaces and solid elements. Therefore, it can precisely represent CAD model.

However, several disadvantages of the IGES standard in relation to its use as a RP format include the following objections:

(1) Because IGES is the standard format to exchange data between CAD systems, it also includes much redundant information that is not needed for rapid prototyping systems.
(2) The algorithms for slicing an IGES file are more complex than the algorithms slicing a STL file.
(3) The support structures needed in RP systems such as the SLA cannot be created according to the IGES format.

IGES is a generally used data transfer medium which interfaces with various CAD systems. It can precisely represent a CAD model. Advantages of using IGES over current approximate methods include precise geometry representations, few data conversions, smaller data files and simpler control strategies. However, the problems are the lack of transfer standards for a variety of CAD systems and system complexities.

6.5.2 HP/GL File

HP/GL (Hewlett-Packard Graphics Language) is a standard data format for graphic plotters [1, 2]. Data types are all two-dimensional, including lines, circles, splines, texts, etc. The approach, as seen from a designer's point of view, would be to automate a slicing routine which generates a section slice, invoke the plotter routine to produce a plotter output file and then loop back to repeat the process.

The advantages of the HP/GL format are that a lot of commercial CAD systems have the interface to output the HP/GL format and it is a 2D geometry data format which does not need to be sliced.

However, there are two distinct disadvantages of the HP/GL format. First, because HP/GL is a 2D data format, the files would not be appended, potentially leaving hundreds of small files needing to be given logical names and then transferred. Second, all the support structures required must be generated in the CAD system and sliced in the same way.

6.5.3   CT Data

CT (Computerized Tomography) scan data is a particular approach for medical imaging [1, 14]. This is not standardized data. Formats are proprietary and somewhat unique from one CT scan machine to another. The scan generates data as a grid of three-dimensional points, where each point has a varying shade of gray indicating the density of the body tissue found at that particular point. Data from CT scans have been used to build skull, femur, knee, and other bone models on Stereolithography systems. Some of the reproductions were used to generate implants, which have been successfully installed in patients. The CT data consist essentially of raster images of the physical objects being imaged. It is used to produce models of human temporal bones.

There are three approaches to making models out of CT scan information: (1) Via CAD Systems (2) STL-interfacing and (3) Direct Interfacing. The main advantage of using CT data as an interface of rapid prototyping is that it is possible to produce structures of the human body by the rapid prototyping systems. But, disadvantages of CT data include firstly, the increased difficulty in dealing with image data as compared with STL data and secondly, the need for a special interpreter to process CT data.

6.6   NEWLY PROPOSED FORMATS

As seen above, the STL file — a collection of coordinate values of triangles — is not ideal and has inherent problems in this format. As a result, researchers including the inventor of STL, 3D Systems Inc., USA, have in recent years proposed several new formats and these are discussed in the following sections. However, none of these has been accepted yet as a replacement of STL. STL files are still widely used today.

6.6.1   SLC File

The SLC (StereoLithography Contour) file format is developed at 3D Systems, USA [15]. It addresses a number of problems associated with the STL format. An STL file is a triangular surface representation of a CAD model. Since the CAD data must be translated to this faceted representation, the surface of the STL file is only an approximation of the real surface of an object. The facets created by STL translation are sometimes noticeable on rapid prototyping parts (such as the AutoCAD Designer part). When the number of STL triangles is increased to produce smoother part surfaces, STL files become very large and the time required for a rapid prototyping system to calculate the slices can increase.

SLC attempts to solve these problems by taking two-dimensional slices directly from a CAD model instead of using an intermediate tessellated STL model. According to 3D Systems, these slices eliminate the facets associated with STL files because they approximate the contours of the actual geometry.

Three problems may arise from this new approach. Firstly, in slicing a CAD model, it is not always necessarily more accurate as the contours of each slice are still approximations of the geometry. Secondly, slicing in this manner requires much more complicated calculations (and therefore, is very time-consuming) when compared to the relatively straightforward STL files. Thirdly, a feature of a CAD model which falls between two slices, but is just under the tolerances set for inclusion on either of the adjacent slices, may simply disappear.

6.6.1.1 *SLC File Specification*

The SLC file format is a "21/2D" contour representation of a CAD model. It consists of successive cross-sections taken at ascending Z intervals in which solid material is represented by interior and exterior boundary polylines. SLC data can be generated from various sources, either by conversion from CAD solid or surface models, or more directly from systems which produce data arranged in layers, such as CT-scanners.

6.6.1.2 *Definition of Terms*

6.6.1.2.1 Segment

A segment is a straight line connecting two *X/Y* vertice points.

6.6.1.2.2 Polyline

A polyline is an ordered list of *X/Y* vertice points connected continuously by each successive line segment. The polyline must be closed whereby the last point must equal the first point in the vertice list.

6.6.1.2.3 Contour boundary

A boundary is a closed polyline representing interior or exterior solid material. An exterior boundary has its polyline list in counter-clockwise order. The solid material is inside the polyline. An interior boundary has its polyline list in clockwise order and solid material is outside the polyline. Figure 6.28 shows a description of the contour boundary.

6.6.1.2.4 Contour layer

A contour layer is a list of exterior and interior boundaries representing the solid material at a specified Z cross-section of the CAD model. The cross-section slice is taken parallel to the *X/Y* plane and has a specified layer thickness.

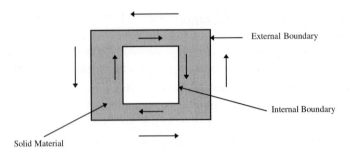

Figure 6.28: Contour boundary description

6.6.1.3 Data Formats

Byte	8 bits
Character	1 Byte
Unsigned Integer	4 Bytes
Float	4 Bytes IEEE Format

The most significant byte of FLOAT is specified in the highest addressed byte. The byte ordering follows the Intel PC Little Indian/Big Indian scheme.

Address	0	1	2	3
	Low Word		High Word	
	LSB MSB		LSB MSB	

Most UNIX RISC Workstations are Big Indian/Little Indian, therefore they need to byte swap all Unsigned Integers and Floats before outputting to the SLC file.

6.6.1.4 Overview of the SLC File Structure

The SLC file is divided into a header section, a 3D reserved section, a sample table section, and the contour data section.

6.6.1.4.1 Header section

The Header section is an ASCII character string containing global information about the part and how it was prepared.

The header is terminated by a carrage return, line feed and control-Z character (0x0d,0x0a,0x1a) and can be a maximum of 2048 bytes including the termination characters.

The syntax of the header section is a keyword followed by its appropriate parameter.

6.6.1.4.2 Header keywords

- "-SLCVER <X.X>" specifies the SLC file format version number. The version number of this specification is 2.0.
- "-UNIT <INCH/MM>" indicates which units the SLC data is represented.
- "-TYPE <PART/SUPPORT/WEB>" specifies the CAD model type. PART and SUPPORT must be closed contours. WEB types can be open polylines or line segments.
- "-PACKAGE <vendor specific>" identifies the vendor package and version number which produced the SLC file. A maximum of 32 bytes.
- "-EXTENTS <minx,maxx miny,maxy minz,maxz>" describes the X, Y, and Z extents of the CAD model.
- "-CHORDDEV <value>" specifies the cordal deviation, if used, to generate the SLC data.
- "-ARCRES <value in degrees>" specifies the arc resolution, if used, to generate the SLC data.
- "-SURFTOL <value>" specifies the surface tolerance, if used, to generate the SLC data.
- "-GAPTOL <value>" specifies the gap tolerance, if used, to generate the SLC data.
- "-MAXGAPFOUND <value>" specifies the maximum gap size found when generating the SLC data.
- "-EXTLWC <value>" specifies, if any, line width compensation has been applied to the SLC data by the CAD vendor.

6.6.1.4.3 3D reserved section

This 256 byte section is reserved for future use.

6.6.1.4.4   Sampling table section

The sample table describes the sampling thicknesses (layer thickness or slice thickness) of the part. There can be up to 256 entries in the table. Each entry descibes the Z start, the slice thickness, and what line width compensation is desired for that sampling range.

Sampling Table Size	1 Byte
Sampling Table Entry	4 Floats
Minimum Z Level	1 Float
Layer Thickness	1 Float
Line Width Compensation	1 Float
Reserved	1 Float

The first sampling table entry Z start value must be the very first Z contour layer. For example, if the cross-sections were produced with a single thickness of 0.006 inches and the first Z level of the part is 0.4 inches and a line width compensation value of 0.005 is desired, then the sampling table will look like the following:

Sample Table Size	1
Sample Table Entry	0.4 0.006 0.005 0.0

If for example, the part was sliced with two different layer thicknesses, the sample table could look like the following:

Sample Table Size	2
Sample Table Entry 1	0.4 0.005 0.004 0.0
Sample Table Entry 2	2.0 0.010 0.005 0.0

Slice thicknesses must be even multiples of one other to avoid processing problems.

6.6.1.4.5   Contour data section

The contour data section is a series of successive ascending Z cross-sections or layers with the accompanying contour data. Each contour layer contains the minimum Z layer value, number of boundaries followed by the list of individual boundary data. The boundary data contains the number of x, y vertices for that boundary, the number of gaps, and finally the list of floating point vertice points.

The location of a gap can be determined when a vertice point repeats itself.

To illustrate, given the contour layer in Section 2.4 the contour section could be as follows:

Z Layer	0.4
Number of Boundaries	2
Number of Vertices for the 1st Boundary	5
Number of Gaps for the 1st Boundary	0
Vertex List for 1st Boundary	0.0, 0.0
	1.0, 0.0
	1.0, 1.0
	0.0, 1.0
	0.0, 0.0

*Notice the direction of the vertice list is counter-clockwise indicating that the solid material is inside the polylist. Also, notice that the polylist is closed because the last vertice is equal to the first vertice.

Number of Vertices for the 2nd Boundary	5
Number of Gaps for the 2nd Boundary	0
Vertex List for 2nd Boundary	0.2, 0.2
	0.2, 0.8
	0.8, 0.8
	0.8, 0.2
	0.2, 0.2

*Notice the direction of the vertice list is clockwise indicating the solid material is outside the polylist. Also, notice that the polylist is closed because the last vertice is equal to the first vertice.

The contour layers are stacked in ascending order until the top of the part. The last layer or the top of the part is indicated by the Z level and a termination unsigned integer (0xFFFFFFFF).

Contour Layer Section Description

Contour Layer

Minimum Z Level	Float
Number of Boundaries	Unsigned Integer
Number of Vertices	Unsigned Integer
Number of Gaps	Unsigned Integer
Vertices List (*X/Y*)	Number of Vertices * 2 Float
Repeat Number of Boundaries	1

Repeat Contour Layer until Top of Part

Top of Part

Maximum Z Level	1 Float
Termination Value	Unsigned Integer (0xFFFFFFFF)

Minimum Z Level for a Given Contour Layer

A one inch cube based at the origin 0, 0, 0 can be represented by only one contour layer and the Top of Part Layer data.

Suppose the cube was to be imaged in 0.010 layers. The sample table would have a single entry with its starting Z level at 0.0 and layer thickness at 0.01. The contour layer data section could be as follows:

Z Layer	0.0
Number of Boundaries	1
Number of Vertices for the 1st Boundary	5
Number of Gaps for the 1st Boundary	0
Vertex List for 1st Boundary	0.0, 0.0
	1.0, 0.0
	1.0, 1.0
	0.0, 1.0
	0.0, 0.0
Z Layer	1.0
Termination Value	0xFFFFFFFF

Notice, only one contour was necessary to describe the entire part. The initial contour will be imaged until the next minimum contour layer or the top of the part at the specified layer thickness described in the sampling table. Now, this part could have 100 identical contour layers, but that would have been redundant. This is why the contour Z value is referred to as the minimum Z value. It gets repeated until the next contour or top of the part.

6.6.2   CLI File

The CLI (Common Layer Interface) format is developed in a Brite Euram project [2, 16] with the support of major European car manufacturers. The CLI format is meant as a vendor-independent format for layer by layer manufacturing technologies. In this format, a part is built by a succession of layer descriptions. The CLI file can be in binary or ASCII format. The geometry part of the file is organized in layers in the ascending order. Every layer is started by a layer command, giving the height of the layer.

The layers consist of series of geometric commands. The CLI format has two kinds of entities. One is the polyline. The polylines are closed, which means that they have a unique sense, either clockwise or anti-clockwise. This directional sense is used in the CLI format to state whether a polyline is on the outside of the part or surrounding a hole in the part. Counter-clockwise polylines surround the part, whereas clockwise polylines surround holes. This allows correct directions for beam offset.

The other is the hatching to distinguish between the inside and outside of the part. As this information is already present in the direction of polyline, and hatching takes up considerable file space, hatches have not been included into output files.

The advantages of the CLI format are given as follows:

(1) Since the CLI format only supports polyline entities, it is a simpler format compared to the HP/GL format.
(2) The slicing step can be avoided in some applications.
(3) The error in the layer information is much easier to be correct than that in the 3D information. Automated recovery procedures can be used and if required, editing is also not difficult.

However, there exists several disadvantages of the CLI format. They are given as follows:

(1) The CLI format only has the capability of producing polylines of the outline of the slice.
(2) Although the real outline of the part is obtained, by reducing the curve to segments of straight lines, the advantage over the STL format is lost.

The CLI format also includes the layer information like the HP/GL format. But, the CLI format only has polyline entities, while HP/GL supports arcs and lines. The CLI format is simpler than the HP/GL format and has been used by several rapid prototyping systems. It is hoped that the CLI format will become an industrial standard such as STL.

6.6.3   RPI File

The RPI (Rapid Prototyping Interface) format is designed by the Rensselaer Design Research Center [4, 7], Rensselaer Polytechnic Institute. It can be derived from currently accepted STL format data. The RPI format is capable of representing facet solids, but it includes additional information about the facet topology. Topological information is maintained by representing each facet solid entity with indexed lists of vertices, edges, and faces. Instead of explicitly specifying the vertex coordinates for each facet, a facet can refer to them by index numbers. This contributes to the goal of overall redundant information reduction.

The format is developed in ASCII to facilitate cross-platform data exchange and debugging. A RPI format file is composed of the collection of entities, each of which internally defines the data it contains. Each entity conforms to the syntax defined by the syntax diagram shown in Figure 6.29. Each entity is composed of an entity name, a record count, a schema definition, schema termination symbol, and the corresponding data. The data is logically subdivided into records which are made up of fields. Each record corresponds to one variable type in the type definition.

The RPI format includes the following four advantages:

(1) Topological information is added to the RPI format. As the result, flexibility is achieved. It allows users to balance storage and processing costs.
(2) Redundancy in the STL is removed and the size of file is compacted.
(3) Format extensibility is made possible by interleaving the format schema with data as shown in Figure 6.29.
(4) Representation of CSG primitives is provided, as capabilities to represent multiple instances of both facet and CSG solids.

Two disadvantages of the RPI format are given as follows:

(1) An interpreter which processes a format as flexible and extensible as the RPI format, is more complex than that for the STL format.
(2) Surface patches suitable for solid approximation cannot be identified in the RPI format.

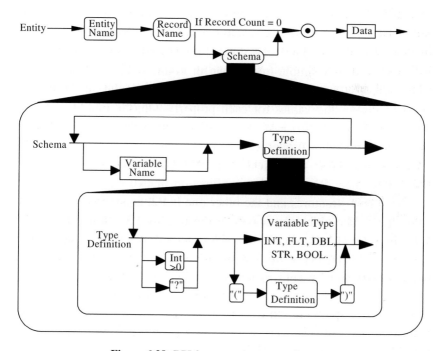

Figure 6.29: RPI format entity syntax diagram

The RPI format offers a number of features unavailable in the STL format. The format can represent CSG primitive models as well as facet models. Both can be operated by the Boolean union, intersection, and difference operators. Provisions for solid translation and multiple instancing are also provided. Process parameters, such as process types, scan methods, materials, and even machine operator instructions, can be included in the file. Facet models are more efficiently represented as redundancy is reduced. The flexible format definition allows storage and processing cost to be balanced.

6.6.4   LEAF File

The LEAF or Layer Exchange ASCII Format, is generated by Helsinki University of Technology [11]. To describe this data model, concepts from the object-oriented paradigm are borrowed. At the top level, there is an object called LMT-file (Layer Manufacture Technology file) that can contain parts which in turn are composed of other parts or by layers. Ultimately, layers are composed of 2D primitives and currently the only ones which are planned for implementation are polylines.

For example, an object of a given class is created. The object classes are organized in a simple tree shown in Figure 6.30. Attached to each object class is a collection of properties. A particular instance of an object specifies the values for each property. Objects inherit properties from their parents. In LEAF, the geometry of an object is simply one among several other properties.

In this example, the **object** is a LMT-file. It contains exactly one child, the object **P1**. **P1** is the combination of two parts, one of which is the support structures and the other one is **P2**, again a combination of two others. The objects at leaves of the tree — **P3**, **P4** and **S** — must have been, evidently, sliced with the same z-values so that the required operations, in this case **or** and **binary-or**, can be performed and the layers of **P1** and **P2** constructed.

Figure 6.30: The object tree

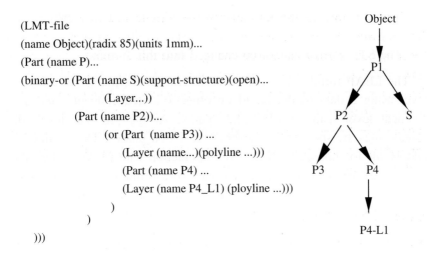

```
(LMT-file
(name Object)(radix 85)(units 1mm)...
(Part (name P)...
(binary-or (Part (name S)(support-structure)(open)...
                (Layer...))
        (Part (name P2))...
            (or (Part (name P3)) ...
                (Layer (name...)(polyline ...)))
                (Part (name P4) ...
                (Layer (name P4_L1) (ployline ...)))
            )
        )
    )))
```

Figure 6.31: An instance tree

In LEAF, the properties support-structure and open can also be attached to layer or even polyline objects allowing the sender to represent the original model and the support structures as one single part. In Figure 6.31, all parts inherit the properties of **object**, their ultimate parent. Likewise, all layers of the object **S** inherit the open property indicating that the contours in the layers are always interpreted as open, even if they are geometrically closed.

Amongst the many advantages of the LEAF format are:

(1) It is easy to implement and use.
(2) It is not ambiguous.
(3) It allows for data compression and for a human-readable representation.
(4) It is machine independent and LMT process independent.
(5) Slices of CSG models can be represented almost directly in LEAF.
(6) The part representing the support structures can be easily separated from the original part.

The disadvantages of the LEAF format include the following items:

(1) The new interpreter is needed for connecting the rapid prototyping systems.

(2) The structure of the format is more complicated than that of the STL format.

(3) The STL format cannot be changed into this format.

The LEAF format is described at several levels, mainly at a logical level using a data model based on object-oriented concepts, and at a physical level using a LISP-like syntax. At the physical level, the syntax rules are specified by several translation phases. Thus defined, it allows one to choose at which level, interaction with LEAF is desirable and at each level there is clear and easy-to-use interface. It is doubtful that LEAF currently supports the needs of all processes currently available but it is hoped it is a step forward in the direction.

6.7   STANDARD FOR REPRESENTING LAYERED MANUFACTURING OBJECTS

Currently, there is the ISO 10303, the international Standard for the Exchange of Product model data (STEP). This standard is intended for the computer-interpretable representation and exchange of product data for engineering purposes [17]. It comprises Generic Resources, Description Methods, Implementation Methods and Application Protocols (APs). However, at present there is no STEP AP that covers layered manufacturing.

The Rapid Prototyping and Layered Manufacturing (RPLM) group is proposing a New Work Item (NWI) for the development of an ISO 10303 AP for RPLM. This AP is to be a standard for product-related data and will support the RPLM process chain. It will comprise the design, process planning, RPLM process, post-processing and measurement of the product life-cycle. These components will require data representations for product geometry, product structure, configuration, tolerance information, product properties (e.g., appearance, color, identification of functional faces) and material data including the inhomogeneous and non-isotropic material distributions that RPLM processes can now generate [18].

Recently, there was a proposed standards-based approach for representing heterogeneous objects for layered manufacturing. The main reason cited was that current solid modeling concentrates on the geometry and topology of the features, but does not include material gradation data [17]. Since layered manufacturing is able to fabricate an object with different materials and compositions, there is an urgent need for a standard to represent heterogeneous objects for layered manufacturing.

The following are several possible representations of the macro-structure of the material for modeling of heterogeneous objects [17]:

(1) tetrahedral decompositions,
(2) voxel-based representation,
(3) R-function method,
(4) r_m object model.

To represent heterogeneous objects in the ISO 10303 domain, a data planning model (DPM) is the required next step. A DPM shows the basic theories of the application domain and the general relationships among the main theories. After drawing up the DPM, the DPM must be validated. This was carried out by physically manufacturing the parts represented in the proposed STEP format. Representation of heterogeneous objects in a standardized format is a vital step in the successful physical realization of heterogeneous objects through layered manufacturing [17].

REFERENCES

[1] Jacobs, P.F., *Rapid Prototyping and Manufacturing*, Society of Manufacturing Engineers, 1992.

[2] Famieson, R. and Hacker, H., "Direct slicing of CAD models for rapid prototyping," *Rapid Prototyping Journal*, ISATA94, Aachen, Germany, 31 Oct to 4 Nov 94.

[3] Donahue, R.J., "CAD model and alternative methods of information transfer for rapid prototyping systems," *Proceedings of the Second International Conference on Rapid Prototyping*, 1991, pp. 217–235.

[4]   Wozny, M.J., "Systems issues in solid freeform fabrication," *Proceedings, Solid Freeform Fabrication Symposium*, 3–5 August 1992, Texas, USA, pp. 1–15.

[5]   Leong, K.F., Chua, C.K., and Ng, Y.M., "A study of stereolithography file errors and repair Part 1 — Generic solutions," *International Journal of Advanced Manufacturing Technologies*, accepted for publication.

[6]   Leong, K.F., Chua, C.K., and Ng, Y.M., "A study of stereolithography file errors and repair Part 2 — Special cases," *International Journal of Advanced Manufacturing Technologies*, accepted for publication.

[7]   Rock, S.J. and Wozny, M.J., "A flexible format for solid freeform fabrication," *Proceedings, Solid Freeform Fabrication Symposium*, 12–14 August 1991, Texas, USA, pp. 1–12.

[8]   Crawford, R.H., "Computer aspects of solid freeform fabrication: Geometry, process control and design," *Proceedings, Solid Freeform Fabrication Symposium*, 9–11 August 1993, Texas, USA, pp. 102–111.

[9]   Bohn, J.H. and Wozny, M.J., "Automatic CAD-model repair: Shell-closure," *Proceedings, Solid Freeform Fabrication Symposium*, 3–5 August 1992, Texas, USA, pp. 86–94.

[10]   Bohn, J.H., *Automatic CAD-Model Repair*, Ann Arbor, Mich., USA, UMI.

[11]   Dolenc, A. and Malela, I., "A data exchange format for LMT processes," *Proceedings of the Third International Conference on Rapid Prototyping*, 1992, pp. 4–12.

[12]   Reed, K., Harrvd, D., and Conroy, W., *Initital Graphics Exchange Specification (IGES)*, Version 5.0, CAD-CAM Data Exchange Technical Centre, 1990.

[13]   Li, Jinghon, "Improving stereolithography parts quality — Practical solutions," *Proceedings of the Third International Conference on Rapid Prototyping*, 1992, pp. 171–179.

[14] Swaelens, B. and Kruth, J.P., "Medical applications of rapid prototyping techniques," *Proceedings of the Fourth International Conference on Rapid Prototyping*, 1993, pp. 107–120.

[15] Vancraen, W., Swawlwns, B., and Pauwels, Johan, "Contour interfacing in rapid prototyping — Tools that make it work," *Proceedings of the Third European Conference on Rapid Prototyping and Manufacturing*, 1994, pp. 25–33.

[16] Smith-Moritz, Geoff, "3D Systems," *Rapid Prototyping Report Rapid Prototyping and Manufacturing* **4**(12) (1994): 3.

[17] Patil, L., Dutta, D., Bhatt, A.D., Jurrens, K., Lyons, K., Pratt, M.J., and Sriram, R.D., "A proposed standards-based approach for representing heterogeneous objects for layered manufacturing," *Rapid Prototyping Journal* **8**(3) (2002): 134–146.

[18] Pratt, M., *New Work Item Proposal*, ISO/TC 184/SC 4 N, June 2000.

PROBLEMS

1. What is the common format used by RP systems? Describe the format and illustrate with an example. What are the pros and cons of using this format?

2. Referring to Figure 6.32, write a sample STL file for the shaded triangle.

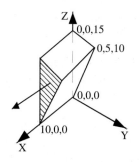

Figure 6.32: Sample STL file

3. Based on the STL format, how many triangles and coordinates would a cube contain?

4. What causes *missing facets or gaps* to occur?

5. Illustrate, with diagrams, the meaning of *degenerate facets*.

6. Explain *overlapping facets*.

7. What are the three types of non-manifold conditions?

8. What are the consequences of building a valid and invalid tessellated model?

9. What problems can the generic solution solve?

10. Describe the algorithm which is used to solve the *missing facets'* problem.

11. For Figure 6.33, facet X is incorrectly orientated. Describe how the problem can be resolved. Draw the newly generated facet X with the corrected orientation.

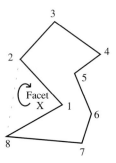

Figure 6.33: Incorrectly generated facet's orientation

12. Describe the algorithm used to solve special case 1 where two or more gaps are formed from a coincidental vertex.

13. Prove and illustrate how a generate facet can be repaired by the use of vector algebra.

14. How can the problem of overlapping facets be solved?

15. What is the efficiency of the detection routine? Illustrate using the example of a cube.

16. What are some of the limitations of the solutions, both generic and special cases, described to solve STL-related problems?

17. Name some other translators used in place of STL.

18. What problems does the SLC file format seek to address?

19. Some newly proposed formats are CLI, RPI and the LEAF files. Describe them briefly and contrast their strengths and weaknesses.

Chapter 7
APPLICATIONS AND EXAMPLES

7.1 APPLICATION-MATERIAL RELATIONSHIP

Areas of applications are closely related to the purposes of prototyping and consequently the materials used. As such, the closer the RP materials to the traditional prototyping materials in physical and behavioral characteristics, the wider will be the range of applications. Unfortunately, there are marked differences in these areas between current RP materials and traditional materials in manufacturing. The key to increasing the applicability of RP technologies therefore lies in widening the range of materials.

In the early developments of RP systems, the emphasis of the tasks at hand was oriented towards the creation of "touch-and-feel" models to support design, i.e., creating 3D objects with little or without regard to their function and performance. These are broadly classified as "Applications in Design". It is the result that influenced, and in many cases limited by, the materials available on these RP systems. However as the initial costs of the machines are high, vendors are constantly in search for more areas of applications, with the logical search for functional evaluation and testing applications, and eventually tooling. This not only calls for improvements in RP technologies in terms of processes to create stronger and more accurate parts, but also in terms of developing an even wider range of materials, including metals and ceramic composites. Applications of RP prototypes were first extended to "Applications in Engineering, Analysis and Planning" and later extended further to "Applications in Manufacturing and Tooling". These typical application areas are summarized in Figure 7.1 and discussed in the following sections.

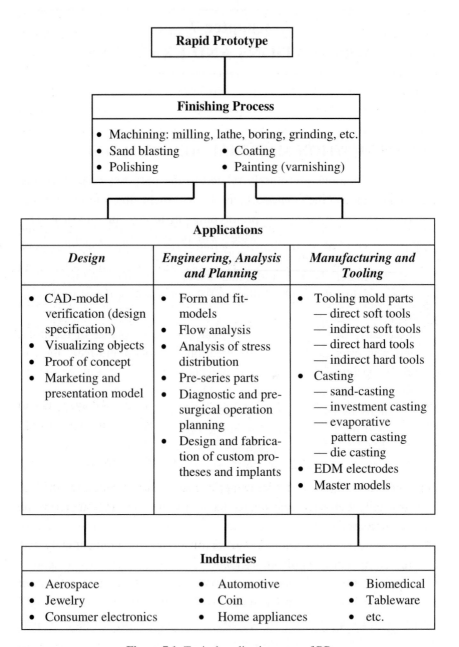

Figure 7.1: Typical application areas of RP

The major breakthrough of RP technologies in manufacturing has been their abilities in enhancing and improving product development while at the same time reducing the costs and time required to take the product from conception to market. Sections 7.6 to 7.11 contain examples of applications in the aerospace, automotive, biomedical, jewelry, coin and tableware industries. These examples are by no means exhaustive, but they do represent their applications in a wide cross-section of the industries.

7.2 FINISHING PROCESSES

As there are various influencing factors such as shrinkage, distortion, curling and accessible surface smoothness, it is necessary to apply some post-RP finishing processes to the parts just after they have been produced. These processes can be carried out before the RP parts are used in their desired applications. Furthermore, additional processes may be necessary in specific cases, e.g., when creating screw threads.

7.2.1 Cutting Processes

In most cases, the resins or other materials used in the RP systems can be subjected to traditional cutting processes, such as milling, boring, turning, and grinding.

These processes are particularly useful for the following:

(1) Deviations in geometrical measurements or tolerances due to unpredictable shrinkage during the curing or bonding stages of the RP process.
(2) Incomplete generation of selected form features. This could be due to fine or complex-shaped features.
(3) Clean removal of necessary support structures or other remainder materials attaching to the RP parts.

In all these cases, it is possible to achieve economic surface finishing of the objects generated with a combination of NC machining and computer-aided NC programming.

7.2.2   Sand-Blasting and Polishing

Sand blasting or abrasive jet deburring can be used as an additional cleaning operation or process to achieve better surface quality. However, there is a trade-off in terms of accuracy. Should better finishing be required, additional polishing by mechanical means with super-fine abrasives can also be used after sandblasting.

7.2.3   Coating

Coating with appropriate surface coatings can be used to further improve the physical properties of the surface of plastic RP parts. One example is galvano-coating, a coating which provides very thin metallic layers to plastic RP parts.

7.2.4   Painting

Painting is applied fairly easily on RP parts made of plastics or paper. It is carried out mainly to improve the aesthetic appeal or for presentation purposes, e.g., for marketing or advertising presentations.

Once the RP parts are appropriately finished, they can then be used for the various areas of application as shown in Figure 7.1.

7.3   APPLICATIONS IN DESIGN

7.3.1   CAD Model Verification

This is the initial objective and strength of RP systems, in that designers often need the physical part to confirm the design that they have created in the CAD system. This is especially important for parts or products designed to fulfill aesthetic functions or that are intricately designed to fulfill functional requirements.

7.3.2   Visualizing Objects

Designs created on CAD systems need to be communicated not only amongst designers within the same team, but also to other departments, like manufacturing, and marketing. Thus, there is a need to create objects from the CAD designs for visualization so that all these people will be referring to the same object in any communications. Tom Mueller in his paper entitled, "Application of Stereolithography in injection molding" [1] characterizes this necessity by saying:

> *"Many people cannot visualize a part by looking at print. Even engineers and toolmakers who deal with print everyday requires several minutes or even hours of studying a print. Unfortunately, many of the people who approve a design (typically senior management, marketing analysts, and customers) have much less ability to understand a design by looking at a drawing."*

7.3.3 Proof of Concept

Proof of concept relates to the adaptation, of specific details to an object environment or aesthetic aspects (such as car telephone in a specific car), or of specific details of the design on the functional performance of a desired task or purpose.

7.3.4 Marketing and Commercial Applications

Frequently, the marketing or commercial departments require a physical model for presentation and evaluation purposes, especially for assessment of the project as a whole. The mock-up or presentation model can even be used to produce promotional brochures and related materials for marketing and advertising even before the actual product becomes available.

7.4 APPLICATIONS IN ENGINEERING, ANALYSIS AND PLANNING

Other than creating a physical model for visualization or proofing purposes, designers are also interested in the engineering aspects of their designs. This invariably relates to the functions of the design. RP technologies become important as they are able to provide the information necessary to ensure sound engineering and function of the product. What makes it more attractive is that it also save development time and reduce costs. Based on the improved performance of processes and materials available in current RP technologies, some applications for functional models are presented in the following sections.

7.4.1   Scaling

RP technology allows easy scaling down (or up) of the size of a model by scaling the original CAD model. In a case of designing bottles for perfumes with different holding capacities, the designer can simply scale the CAD model appropriately for the desired capacities and view the renderings on the CAD software. With the selected or preferred capacities determined, the CAD data can be changed accordingly to create the corresponding RP model for visualization and verification purposes (see Figure 7.2).

Figure 7.2: Perfume bottles with different capacity

7.4.2   Form and Fit

Other than dealing with sizes and volumes, forms have to be considered from the aesthetics and functional standpoint as well. How a part fits into a design and its environment are important aspects, which have to be addressed. For example, the wing mirror housing for a new car design has to have the form that augments well with the general appearance of the exterior design. This will also include how it fits to the car door. The model will be used to evaluate how it satisfies both aesthetic and functional requirements.

Form and fit models are used not just in the automotive industries. They can also be used for industries involved in aerospace and others like consumer electronic products and appliances.

7.4.3   Flow Analysis

Designs of components that affect or are affected by air or fluid flow cannot be easily modified if produced by the traditional manufacturing routes. However, if the original 3D design data can be stored in a computer model, then any change of object data based on some specific tests can be realized with computer support. The flow dynamics of these products can be computer simulated with software. Experiments with 3D physical models are frequently required to study product performance in air and liquid flow. Such models can be easily built using RP technology. Modifications in design can be done on computer and rebuilt for re-testing very much faster than using traditional prototyping methods. Flow analyses are also useful for studying the inner sections of inlet manifolds, exhaust pipes, replacement heart valves [2], or similar products that at times can have rather complex internal geometries. Should it be required, transparent parts can also be produced using rapid tooling methods to aid visualization of internal flow dynamics. Typically, flow analyses are necessary for products manufactured in the aerospace, automotive, biomedical and shipbuilding industries.

7.4.4   Stress Analysis

In stress analysis using mechanical or photo-optical methods or otherwise, physical replicas of the part being analyzed are necessary. If the material properties or features of the RP technologies generated objects are similar to those of the actual functional parts, they can be used in these analytical methods to determine the stress distribution of the product.

7.4.5   Mock-Up Parts

"Mock-up" parts, a term first introduced in the aircraft industry, are used for final testing of different aspects of the parts. Generally, mock-up parts are assembled into the complete product and functionally tested at pre-determined conditions, e.g., for fatigue. Some RP techniques are able to generate "mock-ups" very quickly to fulfill these functional tests before the design is finalized.

7.4.6   Pre-Production Parts

In cases where mass-production will be introduced once the prototype design has been tested and confirmed, pilot-production runs of ten or more parts is usual. The pilot-production parts are used to confirm tooling design and specifications. The necessary accessory equipment, such as fixtures, chucks, special tools and measurement devices required for the mass-production process are prepared and checked. Many of the RP methods are able to quickly produce pilot-production parts, thus helping to shorten the process development time, thereby accelerating the overall time-to-market process.

7.4.7   Diagnostic and Surgical Operation Planning

In combining engineering prototyping methodologies with surgical procedures, RP models can complement various imaging systems, such as magnetic resonance imaging (MRI) and computed tomography (CT) scanning, to produce anatomical models for diagnostic purposes. These RP models can also be used for surgical and reconstruction operation planning. This is especially useful in surgical procedures that have to be carried out by different teams of medical specialists and where inter-departmental communication is of essence. Several related case studies for these applications can be found in Section 7.8.

7.4.8 Design and Fabrication of Custom Prosthesis and Implant

RP can be applied to the design and fabrication of customized prostheses and implants. A prosthesis or implant can be made from anatomical data inputs from imaging systems, e.g., laser scanning and computed tomography (CT). In cases, such as having to produce ear prostheses, a scan profile can be taken of the good ear to create a computer-mirrored exact replica replacement using RP technology. These models can be further refined and processed to create the actual prostheses or implants to be used directly on a patient. The ability to efficiently customize and produce such prostheses and implants is important, as standard sizes are not always an ideal fit for the patient. Also, a less than ideal fit, especially for artificial joints and weight bearing implants, can often result in accumulative problems and damage to the surrounding tissue structures. Case studies on similar applications can be found in Section 7.8.

7.5 APPLICATIONS IN MANUFACTURING AND TOOLING

Central to the theme of rapid tooling is the ability to produce multiple copies of a prototype with functional material properties in short lead-times. Apart from mechanical properties, the material can also include functionalities such as color dyes, transparency, flexibility and the like. Two issues are to be addressed here: tooling proofs and process planning. Tooling proofs refer to getting the tooling right so that there will not be a need to do a tool change during production because of process problems. Process planning is meant for laying down the process plans for the manufacture as well as assembly of the product based on the prototypes produced.

Rapid tooling can be classified into soft or hard, and direct or indirect tooling [3], as schematically shown in Figure 7.3. Soft tooling, typically made of silicon rubber, epoxy resins, low melting point alloys and foundry sands, generally allows for only single casts or for small

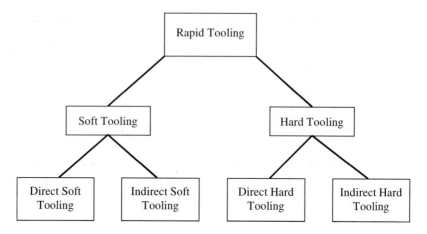

Figure 7.3: Classification of rapid tooling

batch production runs. Hard tooling, on the other hand, usually made from tool steels, generally allows for longer production runs.

Direct tooling is referred to when the tool or die is created directly by the RP process. As an example in the case of injection molding, the main cavity and cores, runner, gating and ejection systems, can be produced directly using the RP process. In indirect tooling, on the other hand, only the master pattern is created using the RP process. A mold, made of silicon rubber, epoxy resin, low melting point metal, or ceramic, is then created from the master pattern.

7.5.1   Direct Soft Tooling

This is where the molding tool is produced directly by the RP systems. Such tooling can be used for liquid metal sand casting, in which the mold is destroyed after a single cast. Other examples, such as composite molds, can be made directly using steoreolithography. These are generally used in the injection molding of plastic components and can withstand up to between 100 to 1000 shots. As these molding tools can typically only support a single cast or small batch production run before breaking down, they are classified as soft tooling. The following section list several examples of direct soft tooling methods.

7.5.1.1 *Selective Laser Sintering® of Sand Casting Molds*

Sand casting molds can be produced directly using the selective laser sintering (SLS®) process. Individual sand grains are coated with a polymeric binder. Laser energy is applied to melt this binder which coats the individual sand grains together, thereby bonding the grains of sand together in the shape of a mold [4]. Accuracy and surface finish of the metal castings produced from such molds are similar to those produced by conventional sand casting methods. Functional prototypes can be produced this way, and should modifications be necessary, a new prototype can be produced within a few days.

7.5.1.2 *Direct AIM*

A rapid tooling method developed by 3D CAD/CAM systems uses the SLA to produce resin molds that allow the direct injection of thermoplastic materials. Known as the Direct AIM (ACES injection molding) [5], this method is able to produce high levels of accuracy. However, build times using this method are relatively slow on the standard stereolithography (SLA) machine. Also, because the mechanical properties of these molds are very low, tool damage can occur during ejection of the part. This is more evident when producing geometrically more complex parts using these molds.

7.5.1.3 *SL Composite Tooling*

This method builds molds with thin shells of resin with the required surface geometry which is then backed-up with aluminum powder-filled epoxy resin to form the rest of the mold tooling [6]. This method is advantageous in that higher mold strengths can be achieved when compared to those produced by the Direct AIM method which builds a solid SLA resin mold. To further improve the thermal conductivity of the mold, aluminum shot can be added to back the thin shell, thus promoting faster build times for the mold tooling. Other advantages of this method include higher thermal conductivity of the mold and lower tool development costs when compared to molds produced by the Direct AIM method.

7.5.2   Indirect Soft Tooling

In this rapid tooling method, a master pattern is first produced using RP. From the master pattern, a mold tooling can be built out of an array of materials such as silicon rubber, epoxy resin, low melting point metals, and ceramics.

7.5.2.1   *Arc Spray Metal Tooling*

Using metal spraying on the RP model, it is possible to create very quickly an injection mold that can be used to mold a limited number of prototype parts. The metal spraying process is operated manually, with a hand-held gun. An electric arc is introduced between two wires, which melts the wires into tiny droplets [7]. Compressed air blows out the droplets in small layers of approximately 0.5 mm of metal.

The master pattern produced by any RP process is mounted onto a base and bolster, which are then layered with a release agent. A coating of metal particles using the arc spray is then applied to the master pattern to produce the female form cavity of the desired tool. Depending on the type of tooling application, a reinforcement backing is selected and applied to the shell. Types of backing materials include filled epoxy resins, low-melting point metal alloys and ceramics. This method of producing soft tooling is cost and lead-time saving. A typical metal spray process for creating an injection mold is shown in Figure 7.4.

7.5.2.2   *Silicon Rubber Molds*

In manufacturing functional plastic, metal and ceramic components, vacuum casting with the silicon rubber mold has been the most flexible rapid tooling process and the most used to date. They have the following advantages:

• Extremely high resolution of master model details can be easily copied to the silicon cavity mold.

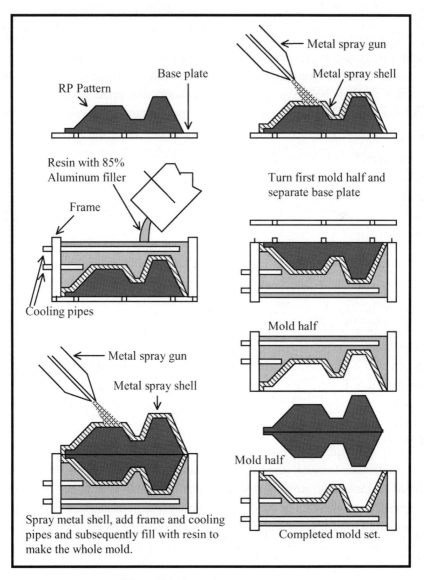

Figure 7.4: A metal arc spray system

- Gross reduction of backdraft problems (i.e., die lock, or the inability to release the part from the mold cavity because some of the geometry is not within the same draw direction as for the rest of the part).

The master pattern, attached with a system of sprue, runner, gating and air vents, is suspended in a container. Silicon rubber slurry is poured into the container engulfing the master pattern. The silicon rubber slurry is baked at 70°C for three hours and upon solidification, a parting line is cut with a scalpel. The master pattern is removed from the mold thus forming the tool cavity. The halves of the mold are then firmly taped together. Materials, such as polyurethane, are poured into the silicon tool cavity under vacuum to avoid asperities caused by entrapped air. Further baking at 70°C for four hours is carried out to cure the cast polymer part. The vacuum casting process is generally used with such molds. Each silicon rubber mold can produce up to 20 polyurethane parts before it begins to break apart [8]. These problems are commonly encountered when using hard molds, making it necessary to have expensive inserts and slides. They can be cumbersome and take a longer time to produce. These are virtually eliminated when the silicon molding process is used.

RP models can be used as master patterns for creating these silicon rubber molds. Figures 7.5(a)–7.5(f) describe the typical process of creating a silicon rubber mold and the subsequent urethane-based part.

A variant of this is a process developed by Shonan Design Co. Ltd. This process, referred to as the "Temp-less" (temperature-less) process, makes use of similar principles in preparing the silicon mold and casting the liquid polymer except that no baking is necessary to cure the materials. Instead, ultraviolet rays are used for curing of the silicon mold and urethane parts. The advantages this gives is a higher accuracy in replicating the master model because no heat is used, less equipment is required, and it takes only about 30% of the time to produce the parts as compared to the standard silicon molding processes [9].

7.5.2.3 *Spin Casting with Vulcanized Rubber Molds*

Spin casting, as its name implies, applies spinning techniques to produce sufficient centrifugal forces in order to assist in filling the

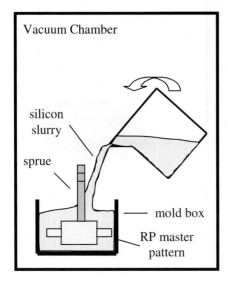

(a) Producing the silicon mold

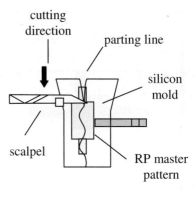

(b) Removing the RP master pattern

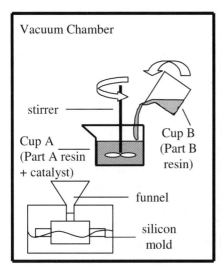

(c) Mixing the resin and catalyst

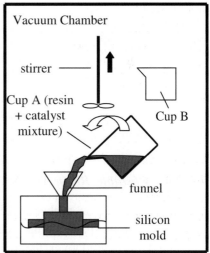

(d) Casting the polymer mixture

Figure 7.5(a)–(f): Vacuum casting with silicon molding

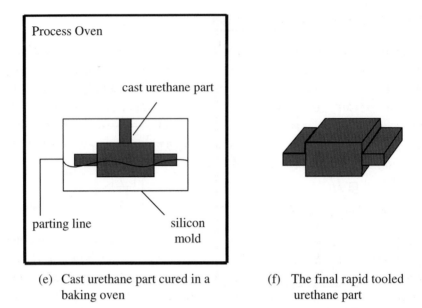

(e) Cast urethane part cured in a
     baking oven

(f) The final rapid tooled
     urethane part

Figure 7.5(a)–(f): (*Continued*)

cavities. Circular tooling molds made from vulcanized rubber are produced in much the same way as in silicon rubber molding. The tooling cavities are formed closer to the outer parameter of the circular mold to increase centrifugal forces. Polyurethane or zinc-based alloys can be cast using this method [10]. This process is particularly suitable for producing low volumes of small zinc prototypes that will ultimately be mass-produced by die-casting.

7.5.2.4   *Castable Resin Molds*

Similar to the silicon rubber molds, the master pattern is placed in a mold box with the parting, line marked out in plasticine [11]. The resin is painted or poured over the master pattern until there is sufficient material for one half of the mold. Different tooling resins may be blended with aluminum powder or pellets so as to provide different mechanical and thermal properties. Such tools are able to withstand up to between 100 to 200 injection molding shots.

7.5.2.5 Castable Ceramic Molds

Ceramic materials that are primarily sand-based can be poured over a master pattern to create the mold [12]. The binder systems can vary with the preference of binding properties. For example, in colloidal silicate binders, the water content in the system can be altered to improve shrinkage and castability properties. The ceramic-binder mix can be poured under vacuum conditions and vibrated to improve the packing of the material around the master pattern.

7.5.2.6 Plaster Molds

Casting into plaster molds has been used to produce functional prototypes [13]. A silicon rubber mold is first created from the master pattern and a plaster mold is then made from this. Molten metal is then poured into the plaster mold which is broken away once the metal has solidified. Silicon rubber is used as an intermediate stage because the pattern can be easily separated from the plaster mold.

7.5.2.7 Casting

In the metal casting process, a metal, usually an alloy, is heated until it is in a molten state, whereupon it is poured into a mold or die that contains a cavity. The cavity will contain the shape of the component or casting to be produced. Although there are numerous casting techniques available, three main processes are discussed here: the conventional sand casting, investment casting, and evaporative casting processes. RP models render themselves well to be the master patterns for the creation of these metal dies.

Sand casting molds are similarly created using RP master patterns. RP patterns are first created and placed appropriately in the sand box. Casting sand is then poured and packed very compactly over the pattern. The box (cope and drag) is then separated and the pattern carefully removed leaving behind the cavity. The box is assembled together again and molten metal is cast into the sand mold. Sand casting is the cheapest and most practical method for the casting of

Figure 7.6: Cast metal (left) and RP pattern for sand casting (Courtesy of Helysis Inc.)

large parts. Figure 7.6 shows a cast metal mold resulting from a RP pattern.

Another casting method, the investment casting process, is probably the most important molding process for casting metal. Investment casting molds can be made from RP pattern masters. The pattern is usually wax, foam, paper or other materials that can be easily melted or vaporized. The pattern is dipped in a slurry of ceramic compounds to form a coating, or investment shell, over it [14]. This is repeated until the shell builds up thickness and strength. The shell is then used for casting, with the pattern being melted away or burned out of the shell, resulting in a ceramic cavity. Molten metal can then be poured into the mold to form the object. The shell is then cracked open to release the desired object in the mold. The investment casting process is ideal for casting miniature parts with thin sections and complex features. Figure 7.7 schematically shows the investment casting process from a RP-produced wax master pattern while Figure 7.8 shows an investment casting mold resulting from a RP pattern.

The third casting process discussed in this book is the evaporative pattern casting. As its names implies, it uses an evaporative pattern, such as polystyrene foam, as the master pattern. This pattern can be produced using the selective laser sintering (SLS) process along with the CastForm™ polystyrene material. The master pattern is attached to

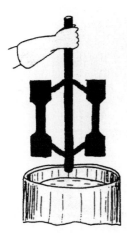

(a) Pattern clusters are dipped in ceramic slurry.

(b) Refractory grain is sifted onto the coated patterns. Steps (a) and (b) are repeated several times to obtain desired shell thickness.

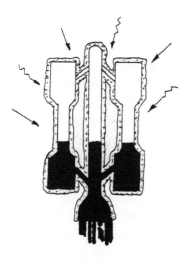

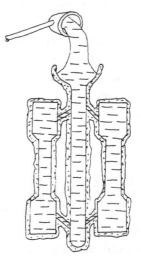

(c) After the mold material has set and dried, the patterns are melted out of the mold.

(d) Hot molds are filled with metal by gravity, pressure vacuum, or centrifugal force.

Figure 7.7: Schematic diagram of the shell investment casting process

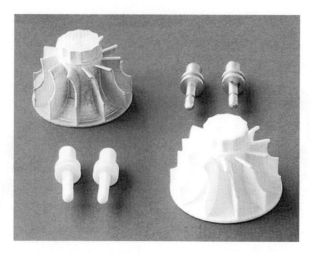

Figure 7.8: Investment casting of fan impeller from RP pattern

sprue, riser and gating systems to form a "tree". This polystyrene "tree" is then surrounded by foundry sand in a container and vacuum compacted to form a mold. Molten steel is then poured into the container through the sprue. As the metal fills the cavity, the polystyrene evaporates with a very low ash content [15]. The part is cooled before the casting is removed. A variety of metals, such as titanium, steel, aluminum, magnesium and zinc can be cast using this method. Figure 7.9 shows schematically how an RP master pattern is used with the evaporative pattern casting process.

7.5.3   Direct Hard Tooling

Hard tooling produced by RP systems has been a major topic for research in recent years. Although several methods have been demonstrated, much research is still being carried out in this area. The advantages of hard tooling produced by RP methods are fast turnaround times to create highly complex-shaped mold tooling for high volume production. The fast response to modifications in generic designs can be almost immediate. The following are some examples of direct hard tooling methods.

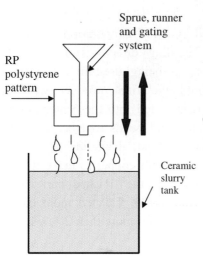

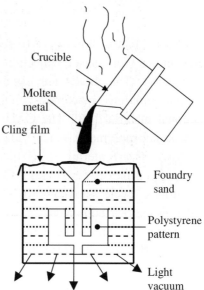

(a) Polystyrene RP pattern "tree" is coated by dipping with a ceramic slurry and air-dried.

(b) Coated RP pattern is packed with foundry sand in a container. The container is sealed with cling film and vacuumed to compact the sand further.

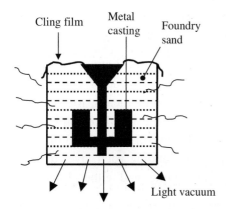

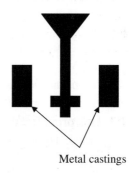

(c) The polystyrene pattern evaporates as the molten metal is cast into the mold. The casting is then left to cool.

(d) After solidification, the final cast parts are removed from the sprue, runner and gating system.

Figure 7.9: Evaporation pattern casting process

7.5.3.1   *RapidTool*™

RapidTool™ is a technology invented by DTM Corporation to produce metal molds for plastic injection molding directly from the SLS Sinterstation. The molds are capable of being used in conventional injection molding machines to mold the final product with the functional material [16]. The CAD data is fed into the Sinterstation™ which bonds polymeric binder coated metal beads together using the Selective Laser Sintering (SLS) process. Next, debinding takes place and the green part is cured and infiltrated with copper to make it solid. The furnace cycle is about 40 hours with the finished part having similar properties equivalent to aluminum. The finished mold can be easily machined. Shrinkage is reported to be no more than 2%, which is compensated for in the software.

Typical time frames allow relatively complex molds to be produced in two weeks as compared to 6 to 12 weeks using conventional techniques. The finished mold is capable of producing up to tens of thousands injection-molded parts before breaking down.

7.5.3.2   *Laminated Metal Tooling*

This is another method that may prove promising for RT applications. The process applies metal laminated sheets with the Laminated Object Manufacturing (LOM) method. The sheets can be made of steel or any other material which can be cut by the appropriate means, for example by CO_2 laser, water jet, or milling, based on the LOM principle [17]. The CAD 3D data provides the sliced 2D information for cutting the sheets layer by layer. However, instead of bonding each layer as it is cut, the layers are all assembled after cutting and either bolted or bonded together.

7.5.3.3   *Direct Metal Laser Sintering (DMLS) Tooling*

The Direct Metal Laser Sintering (DMLS) technology was developed by EOS. The process uses a very high-powered laser to sinter metal powders directly. The powders available for use by this technology are the bronze-based and steel-based materials. Bronze is used for applications where strength requirements are not crucial. Upon sintering of the bronze powder, an organic resin, such as epoxy, is used to infiltrate the part. For steel powders, the process is capable of producing direct steel parts of up to 95% density so that further infiltration is not required. Several direct applications produced with this technology including mold inserts and other metal parts [18].

7.5.3.4   *ProMetal*™ *Rapid Tooling*

Based on MIT's Three Dimensional Printing (3DP) process, the ProMetal™ Rapid Tooling System is capable of creating steel parts for tooling of plastic injection molding parts, lost foam patterns and vacuum forming. This technology uses an electrostatic ink jet print head to eject liquid binders onto the powder, selectively hardening slices of an object a layer at a time. A fresh coat of metal powder is spread on top and the process repeats until the part is completed. The loose powder act as supports for the object to be built. The RP part is then infiltrated at furnace temperatures with a secondary metal to achieve full density. Toolings produced by this technology for use in injection molding have reported withstanding pressures up to 30 000 psi (200 MPa) and surviving 100 000 shots of glass-filled nylon [19].

7.5.4   Indirect Hard Tooling

There are numerous indirect RP tooling methods that fall under this category and this number continues to grow. However, many of these processes remain largely similar in nature except for small differences, e.g., binder system formulations or type of system used. Processes include the Rapid Solidification Process (RSP), Ford's (UK) Sprayform, Cast Kirksite Tooling, CEMCOM's Chemically Bonded Ceramics (CBC) and Swift Technologies Ltd. "SwiftTool", just to name a few. This section will only cover selected processes that can also be said to generalize all the other methods under this category. In general, indirect methods for producing hard tools for plastic injection molding generally make use of casting of liquid metals or steel powders in a binder system. For the latter, debinding, sintering and infiltration with a secondary material are usually carried out as post-processes.

7.5.4.1 *3D Keltool*

The 3D Keltool process has been developed by 3D Systems to produce a mold in fused powdered steel [20]. The process uses a SLA model of the tool for the final part that is finished to a high quality by sanding and polishing. The model is placed in a container where silicon rubber is poured around it to make a soft silicon rubber mold that replicates the female cavity of the SLA model. This is then placed in a box and then silicon rubber is poured around it to produce a replica copy of the SLA model in silicon rubber. This silicon rubber is then placed in a box and a proprietary mixture of metal particles, such as tool steel, and a binder material is poured around it, cured and separated from the silicon rubber model. This is then fired to eliminate the binder and sinter the green metal particles together. The sintered part which is about 70% steel and 30% void is then infiltrated with copper to give a solid mold, which can be used in injection molding.

An alternative to this process is described as the reverse generation process. This uses a positive SLA master pattern of the mold and requires one step less. This process claims that the CAD solid model to injection-molded production part can be completed in four to six weeks. Cost savings of around 25% to 40% can be achieved when compared to that of conventional machined steel tools.

7.5.4.2 *EDM Electrodes*

A method successfully tested in research laboratories but so far not widely applied in industry is the possible manufacturing of copper electrodes for EDM (Electro-Discharge Machining) processes using RP technology. To create the electrode, the RP-created part is used to create a master for the electrode. An abrading die is created from the master by making a cast using an epoxy resin with an abrasive component. The resulting die is then used to abrade the electrode. A specific advantage of the SLS procedure (see Section 5.1) is the possible usage of other materials. Using copper in the SLS process, it is possible to quickly and affordably generate the electrodes used in electrode EDM.

7.5.4.3 *Ecotool*

This is a development between the Danish Technological Institute (DTI) in Copenhagen, Denmark, and the TNO Institute of Industrial Technology of Delft in Holland. The process uses a new type of powder material with a binder system to rapidly produce tools from RP models. However, as its name implies, the binder is friendly to the environment in that it uses a water-soluble base. An RP master pattern is used and a parting line block is produced. The metal powder-binder mixture is then poured over the pattern and parting block and left to cure for an hour at room temperature. The process is repeated to produce the second half of the mold in the same way. The pattern is then removed and the mold baked in a microwave oven.

7.5.4.4 *Copy Milling*

Although not broadly applied nowadays, RP master patterns can be provided by manufacturers to their vendors for use in copy milling, especially if the vendor for the required parts is small and does not have the more expensive but accurate CNC machines. In addition, the principle of generating master models only when necessary, allows some storage space to be saved. The limitation of this process is that only simple geometrical shapes can be made.

7.6 AEROSPACE INDUSTRY

With the various advantages that RP technologies promise, it is only natural that high value-added industries like the aerospace industry have taken special interest in it even though initial investment costs may be high. There are abundant examples of the use of RP technology in the aerospace industry. The following are a few examples.

7.6.1   Design Verification of an Airline Electrical Generator

Sundstrand Aerospace, which manufactures inline electrical generators for military and commercial aircraft, needed to verify its design of an integrated drive generator for a large jetliner [21]. It decided to use Helisys's LOM to create the design-verification model. The generator is made up of an external housing and about 1200 internal parts. Each half of the housing measures about 610 mm in diameter and 300 mm tall and has many intricate internal cavities into which the sub-assemblies must fit.

Such complex designs are difficult to visualize from two-dimensional drawings. A physical model of the generator housing and many of its internal components is a good way to identify design problems before the expensive tooling process. But the time and expense needed to construct the models by traditional means are prohibitive. Thus Sundstrand decided to turn to RP technologies. Initial designs for the generator housing and internal sub-assemblies were completed on a CAD system and the subsequent STL files were sent to a service bureau. Within two weeks, Sundstrand was able to receive the parts from the service bureau and began its own design verification.

Sundstrand assembled the various parts and examined them for form, fit, and limit function. Clearances and interferences between the housing and the many sub-assemblies were checked. After the initial inspection, several problematic areas were found which would have otherwise been missed. These were corrected and incorporated into the CAD design, and in some cases, new RP models were made. Apart from design verification, Sundstrand was able to use the physical models to help toolmakers plan and design casting patterns. The models were also used for manufacturing process design, tool checking, and assembly sequence design. Though the approximate cost for the RP models was US$16 500, the savings realized from removing engineering and design changes were immeasurable, and the time saved (estimated to be about eight to ten weeks) was significant.

7.6.2   Engine Components for Fanjet Engine

In an effort to reduce the developmental time of a new engine, AlliedSignal Aerospace used 3D Systems' QuickCast™ to produce a turbofan jet engine for a business aviation jet [22]. Basically, RP is used for the generation of the casting pattern of an impeller compressor shroud engine component. This part is the static component that provides the seal for the high-pressure compressor in the engine. Three different designs were required for testing the cold rig, hot rig and first engine. Using QuickCast™, the 3D Technology Center was able to directly produce patterns for investment castings using the stereolithography technology. The patterns produced were durable, had improved accuracy, good surface finish and were single large piece patterns. In fact, the patterns created were accurate enough that a design revision error in the assembly fixture was easily detected and corrected. With the use of these RP techniques, production time was slashed by eight to ten weeks, and a savings of US$50 000 for tooling in the three design iterations was realized.

7.6.3   Prototyping Air Inlet Housing for Gas Turbine Engine

Sundstrand Power Systems, a manufacturer of auxiliary engines for military and commercial aircraft, needed prototypes of an air inlet housing for a new gas turbine engine [23, 24]. It first needed mock-ups of the complex design, and also several fully functional prototypes to test on the development engines. The part, which measures about 250 mm in height and 300 mm in diameter, has wall thickness as thin as 1.5 mm (see Figure 7.10). It would have been difficult and costly to build using traditional methods.

To realize the part, Sundstrand used DTM's SLS® system (see Section 5.1) at a service bureau to build the evaluation models of the housing and then generate the necessary patterns for investment casting, ultimately the method used for the manufacture. The SLS® system is

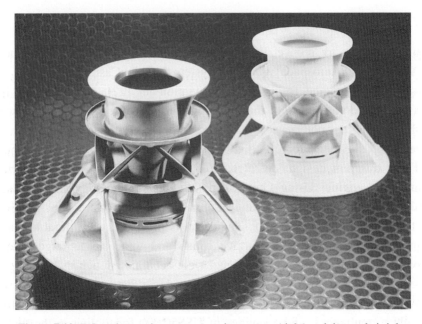

Figure 7.10: Polycarbonate investment-casting pattern (right) and the steel air inlet housing (right) for a jet turbine engine (Courtesy DTM Corporation)

chosen primarily because the air inlet housing has several overhanging structures from which removal of supports would have been extremely difficult.

Sundstrand designed several iterations of the housing as solid models on its CAD system. These models were converted to the STL format and sent to build the nylon evaluation models. As the program progressed, Sundstrand wanted to test the part. As the designs were finalized, new SLS® versions of the part were created as tooling for investment casting. Polycarbonate patterns were created, sealed with wax and sent for casting. The patterns were first coated with a thin layer of polyurethane to fill any remaining surface pores and provide the necessary surface finish. Then the patterns were used to cast the part in Iconel 718 steel, which were sent back to Sundstrand for testing. In all, Sundstrand saved more than four months of tooling and prototyping time, and saved more than US$88 000.

7.6.4 Fabrication of Flight-Certified Production Castings

Bell Helicopter has successfully used stereolithography, first to verify parts design, then to aid with fit and functional testing, and finally to produce investment casting patterns for the manufacture of Federal Aviation Authority (FAA)-certifiable production parts [25]. About 50 of the parts that made up the new helicopter's flight control system were developed with stereolithography. The largest support structure for the hydraulic system, measured approximately 500 mm × 500 mm × 200 mm, and the smallest, 25 mm × 25 mm × 1.1 mm. In production, all parts will be investment cast, most in aluminum while others will be in steel alloys.

Initially, half-scale models were used for design verification, as they were large enough to confirm design intent and were much quicker to fabricate on the SLA machines. Once a design was finalized, full-size SLA models were fabricated for use in "virtual installation" [25]. In virtual installation, full-sized SLA parts were assembled with other components and installed on the actual production helicopter in order to test the fit and kinematics of the assembly. Parts used for virtual installation included all the features that would normally be machined into rough production castings. Problems associated with interferences and clearances were identified and rectified before they could arise in later stages, which by then would be more costly to rectify.

After virtual installation, Bell made QuickCast™ investment casting patterns of each part. These patterns were sent for casting, with the resulting parts being sent for FAA flight certification. In previous projects, Bell would have machined parts to simulate production castings and send them for certification. When the castings became available in about 45 weeks, the parts would have to be re-certified. With QuickCast™ patterns, Bell could produce production-grade metal investment castings in as little as three weeks and did not need re-certification when wax tooling eventually becomes available. The overall development time was shortened with the use of SLA models and QuickCast™ for creating investment casting came closed to six months, resulting in substantial cost savings and a better product was offered to the market.

7.7 AUTOMOTIVE INDUSTRY

7.7.1 Prototyping Complex Gearbox Housing for Design Verification

Volkswagen has utilized Helysis's LOM to speed up the development of a large, complex gearbox housing for its Golf and Passat car lines [26]. The CAD model for the housing was extremely complex and difficult to visualize. VW wanted to build a LOM part to check the design of the CAD model and then use the part for packaging studies.

Using traditional methods, such a prototype would be costly and time consuming to build, and it may not be always possible to include all fine details of the design. Fabrication of the model based on drawings was often subjected to human interpretation, and consequently is error-prone, thus further complicating the prototyping process. All these difficulties were avoided by using RP technology as the fabrication of the model was based entirely on the CAD model created.

The gearbox housing was too large for the build volume of the LOM machine. The CAD model was thus split into five sections and re-assembled after fabrication. It took about ten days to make and finish all five sections, and once they were completed, patternmakers glued them together to complete the final model. The LOM model was first used for verifying the design, and subsequently, to develop sand-casting tooling for the creation of metal prototypes. The RP process had shrunk the prototype development time from eight weeks to less than two, and considerable time and cost savings were achieved.

7.7.2 Prototyping Advanced Driver Control System with Stereolithography

At General Motors, in many of its divisions, RP is becoming a necessary tool in the critical race to be first to market [27]. For example, Delco Electronics, its automotive electronics subsidiary, was involved in the development of the Maestro project. Designed to blend an advanced Audio System, a hands-free cellular phone, Global Positioning System (GPS) navigation, Radio Data System (RDS) information, and climate control into a completely integrated driver control system, the Maestro was to be a marvel.

With many uniquely-shaped push-buttons, two active-matrix LCD screens and a local area network allowing for future expansion, the time needed to develop the system was the most critical factor.

Working with Modern Engineering, an engineering service company, Delco Electronics developed the first renderings and concept drawings for the Maestro project. In order to speed up the project, the designers needed the instrument panel with its myriad of push-buttons working early in the design cycle. Unfortunately, the large number of buttons meant a corresponding large number of rubber molds with all the problems associated with the conventional molding process. From the stylist's concepts, models for each button face were manually machined. Once the designs were confirmed, the machined models were laser scanned, generating the CAD data needed for the creation of SLA models. The final prototype buttons needed to be accurate enough to ensure proper fit and function, as well as be translucent, so that they could be back-lit.

The SLA models generated on 3D Systems' SLA machine were accurate enough to be finished, painted and installed in the actual prototype vehicle, eliminating the need for rubber molds. The result was that in less than four months, Delco Electronics was able to complete the functional instrument panel, with all 108 buttons built using the SLA.

7.7.3   Creating Cast Metal Engine Block with RP Process

As new engine design and development is an expensive and time consuming process, the ability to test a new engine and all its auxiliary components before committing to tooling is important in ensuring costs and time savings [28]. The Mercedes-Benz Division of Daimler-Benz AG initiated a program of physical design verification on prototype engines using SLA parts for initial form and fit testing. After initial design reviews, metal components were produced rapidly using the QuickCast™ process.

Their first project was the design and prototyping of a four-cylinder engine block for the new Mercedes-Benz "A-Class" car. The aim was to cast the engine block directly from a stereolithography QuickCast™ pattern. The engine block was designed on Mercedes-Benz own CAD system, and the data were transferred to 3D Systems Technology Center at Darmstadt, where the one-piece pattern of the block was built on the SLA machine. The full scale investment casting pattern was generated in 96 hours.

The pattern was then sent for shell investment casting, resulting in the 300 mm × 330 mm × 457 mm, engine block being cast in A356-T6 aluminum in just five weeks. The completed engine block incorporated the cast-in water jacket, core passage ways, and exhibited Grade B radiographic quality in all areas evaluated. The entire prototyping process using RP technology lasted only six weeks (compared to 15 to 18 weeks using traditional methods), and the approximate cost savings were approximately US$150 000 as compared with traditional methods. These are both significant, especially in the need for a short time-to-market requirement.

7.7.4   Using Stereolithography to Produce Production Tooling

Ford Motor Company has used 3D Systems' QuickCast™ to create the production tool of a rear wiper-motor cover for the 1994 Explorer sport utility vehicle [29, 30]. The part measured approximately 200 mm × 150 mm by 75 mm and was to be injection molded with polypropylene during production. Traditional methods would have provided the necessary tools for molding in three months.

Ford first built the SLA model of the cover and fit it over the wiper motor to verify the design (see Figure 7.11). Dimensional and assembly problems were identified and rectified before the design was confirmed. From the CAD model data, originally created on the CAD software Pro Engineer, the Pro/MOLDDESIGN® software was used to create "negative" mold halves. Shrink factors were then applied to compensate for the photo-curable resin, A2 steel, and polypropylene. The

Figure 7.11: QuickCast™ generated patterns and the investment cast inserts for the rear wiper-motor cover

QuickCast™ process was then used to build the SLA patterns of the actual tool inserts (in halves). They were then investment cast out in A2 tool-steel. Once cast, the tool inserts would be fitted onto an injection molding machine and used to produce the plastic wiper-motor covers. With the application of such "rapid tooling" techniques, Ford was able to start durability and water flow testing eighteen months ahead of schedule, with a cost reduction of 45% and time savings of more than 40%.

7.8 BIOMEDICAL INDUSTRY

From manufacturing of medical devices and creating customized implants and prostheses to surgical planning and education, RP can be applied to enhance medical applications and healthcare delivery. The following sections relate examples of how RP can play a valuable role in the biomedical industry.

7.8.1   Operation Planning for Cancerous Brain Tumor Surgery

In one case study, a patient had a cancerous bone tumor in his temple area and because of that the surgeon would have to access the growth via the front through the right eye socket. The operation was highly dangerous as damage to the brain was likely which would result in the impairment of some motor functions. In any which way, the patient would have lost the function of the right eye [31]. However, before proceeding with the surgery, the surgeon wanted another examination of the tumor location, but this time using a three-dimensional plastic replica of the patient's skull. By studying the model, the surgeon realized that he could re-route his entry through the patient's jawbone, thus avoiding the risk of harming the eye and motor functions. Eventually, the patient lost only one tooth and of course, the tumor. The plastic RP model used by the surgeon was fabricated by the SLA from a series of 2D CT scans of the patient's skull.

This case study is an excellent example of the potential impact of rapid prototyping in the medical arena. Other case studies relating to bone tumors have also been reported to have had successful results [32, 33]. In all cases, not only was the patient spared physical disability and the emotional and financial price tags associated with that, but the surgeon gain an invaluable insight into his patient through the RP model. The value added to the surgeon's pre-surgical planning stage resulted also in a reduced duration of the procedure and thus the risk of infection and operation costs.

7.8.2   Planning Reconstructive Surgery with RP Technology

Due to a traffic accident, a patient had a serious bone fracture on the upper and lateral orbital rim in the skull [34]. In the first reconstructive surgery, the damaged part of the skull was transplanted with the shoulder bone, but shortly after the surgery, the transplanted bone had dissolved. Thus, it was necessary to perform another surgery to transplant an artificial bone that would not dissolve. The conventional procedure of such a surgery would be for surgeons to manually carve the transplanted bone during the operation until it fitted properly. This operation would have required a lot of time, due to the difficulty in carving bone, let alone during the surgery.

Using rapid prototyping, a SLA prototype of the patient's skull was made and then used to prepare an artificial bone that would fit the hole caused by the dissolution. This preparation not only greatly reduced the time required for the surgery, but also improved its accuracy.

In another case involved at Keio University Hospital of Japan [35], a five-month-old baby had a symptom of scaphocephary in the skull. This is a condition that may lead to serious brain damage because it would only permit the skull to grow in the front and rear directions. The procedure required was to take the upper half of the skull apart and reconstruct it completely, so that the skull would not suppress the brain as the baby grows. Careful planning was essential for the success of such a complex operation. Firstly, the prototype of the skull was produced and amputation lines were drawn on the model. Secondly, a surgical rehearsal was first carried out on the model with the amputation of the skull prototype according to the drawn lines followed by the reconstruction of the amputated part. In this case, the rapid prototype of the skull provided the surgical procedure with: (1) a good three-dimensional visualization support for the planning process, (2) an application as training material, and (3) a guide for the real surgery.

7.8.3   Craniofacial Reconstructive Surgery Planning

Restoration of facial anatomy is required in cases of congenital abnormalities, trauma or post cancer reconstruction. In one case, the patient had a deformed jaw by birth, and a surgical operation was necessary to amputate the shorter side of the jaw and change its position [36]. The difficult part of the operation was the evasion of the nerve canal that runs inside the jawbone. Such an operation was impossible in the conventional procedure because there was no way to visualize the inner nerve canal. Using a CAD model reconstructed from the CT images, it clearly showed the position of the canal and simulation of the amputating process on workstations was a good support for surgeons to determine the actual amputation line.

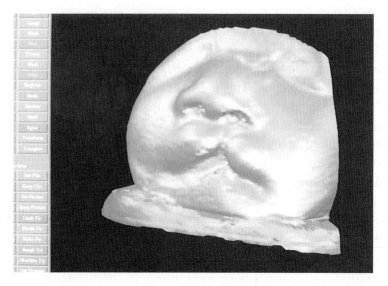

Figure 7.12: CAD model from laser scanner data of a patient's facial details

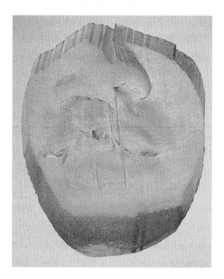

Figure 7.13: SLA model of a patient's facial details

Furthermore, the use of a resin prototype of the jawbone allowed the visualization of the internal nerve canal. The semi-transparent prototype facilitated the determination of the amputation line and enabled an efficient surgery simulation with an actual tool.

In another case study, a laser digitizer was used instead of the CT [37] to capture the facial details of a patient with a harelip problem. The triangulated surface of the patient's face was reconstructed in CAD as seen as Figure 7.12. Figure 7.13 shows the SLA prototype derived from the CAD data. In this case study, the prototype model provided the validation for the laser scan measurements. In addition, it facilitated prediction of the surgical outcome and post-operative assessment of changes in facial surgery. Another advantage was that mirror images could be used to reconstruct the CAD model of facial details such as ears to achieve symmetry.

7.8.4 Biopsy Needle Housing

Biomedical applications are extended beyond design and planning purposes. The prototypes can serve as a master for tooling such as a urethane mold. At Baxter Healthcare, a disposable-medical-products company, designers rely on two RP processes: SLA and SGC to create master models from which they develop metal castings [31]. The masters also serve as a basis for multiple sub-tooling processes. For example, after a master model has been generated via one of the RP machines, the engineers might build a urethane mold around it, cut open the mold, pull out the master, then inject thermoset material into the mold to make the prototype parts.

This process is useful in situations where multiple prototypes are necessary because the engineers can either reuse the rubber molds or make many molds using the same master. The prototypes are then delivered to customer focus groups and medical conferences for professional feedback. Design changes are then incorporated into the master CAD database. Once the design is finalized, the master database is used to drive the machining of the part. Using this method, Baxter Healthcare has made models of biopsy needle housing and many other medical products.

7.8.5   Knee Implants

Engineers at DePuy Inc., a supplier of orthopedic implants, have integrated CAD and RP into their design environment, using it to analyze the potential fit of implants in a specific patient and then modifying the implant design appropriately [31]. At DePuy, SLA plays a major role in the production process of all the company's products, standard and custom. The prototypes are also used as masters for casting patterns to launch a product or to do clinical releases of a product.

For this application, there are several advantages over traditional casting tooling. Firstly, the typical ten-week lead-time is reduced. Secondly, the cost of about US$8000 to US$10 000 is reduced too.

7.8.6   Scaffolds for Tissue Engineering

Tissue engineering has been used to replace failing or malfunctioning organs such as skin, liver, pancreas, heart valve leaflet, ligaments, cartilage and bone. This has given rise to the interests in applying RP techniques to build scaffolds either to induce surrounding tissue and cell in-growth or serve as temporary scaffolds for transplanted cells to attach and grow onto. These scaffolds can be designed in three-dimensions on CAD taking into consideration the porosity and good interconnectivity for tissue induction to occur.

The function of cells, such as in bone and cartilage regeneration, is dependent on the three-dimensional spatial relationships. As such, the geometry of these hard tissues are critical to its function [38–41]. RP has been able to lend itself to producing complex geometry scaffolds.

7.8.7   Customized Tracheobronchial Stents

Stents for maintaining the patency of the respiratory channel has been investigated for production using RP techniques [42]. Customization of these stents can be carried out to take into account compressive resistance with respect to stent wall thickness, as well as unique anatomical considerations. Measurements are taken of the actual forces

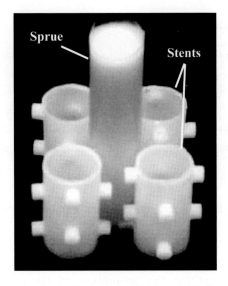

Figure 7.14: Production of the customized stents in slightly differing sizes
for an ideal custom fit

required to open the airway channel to its original dimensions. The data
is fed to the CAD system where modification on the stent design is
carried out. Upon confirmation, the 3D data is fed to an RP system
where the master pattern of the model is built. The master pattern then
undergoes the silicon molding vacuum casting process, reproducing the
stent master pattern with a biocompatible material with all its strength,
spring-back and anti-migration properties in place. Figure 7.14 shows
four vacuum cast tracheobronchial stents in slightly differing sizes for
an ideal intra-surgery custom fit. The stent is sterilized, packaged, and
delivered.

7.8.8   Inter-Vertebral Spacers

Human spinal vertebras can disintegrate due to conditions such as
osteoporosis or extreme forces acting on the spine. In the management
of such situations, a spacer is usually required as part of the spinal
fixation process. RP has been investigated for the production of such

Figure 7.15: Inter-vertebral spacers produced by the RT system at NTU

spacers as it is an ideal process to fabricate 3D structures with good interconnecting pores for the promotion of tissue in-growth. Other considerations for producing such an implant are that the material is biocompatible, and that the mechanical compressive strength of the spacer is able to withstand spinal loads.

A process, developed at the Nanyang Technological University for such a purpose, uses a solid RP master pattern of the spacer to produce a soft mold. Stainless steel bearings coated with a formulated binder system are then cast into the soft mold under vacuum. Upon curing, the part is ejected from the mold. The part then undergoes debinding and sintering processes to produce the final part. The primary advantage of this process is its ability to use a solid RP pattern to produce from this a porous structure with controllable pore sizes and mechanical strengths [43, 44]. Figure 7.15 shows the inter-vertebral spacers using the RT system.

7.8.9   Cranium Implant

A patient suffered from a large frontal cranium defect after complications from a previous meningioma tumor surgery. This left the patient with a missing cranial section, which caused the geometry of the head to look deformed. Conventionally, a titanium-mesh plate would be hand-formed during the operation by the surgeon. This often resulted in inaccuracies and time spent for trial and error. Using RP, standard preparations of the patient were made and a computed tomography (CT) scan of the affected area and surrounding regions was taken during the pre-operation stage. The three-dimensional CT data file was transferred to a CAD system and the missing section of the cranium topography was generated. After some software repair and cleaning up were carried out on the newly generated section, an inverted mold was produced on CAD. This three-dimensional solid model of the mold was saved in *.STL format and transferred to the RP system, such as the SLS®, for building the mold. The SLS® mold was produced and used to mechanically press the titanium-mesh plate to the required three-dimensional profile of the missing cranium section. During the operation, the surgeon cleared the scalp tissue of the defect area and fixated the perfectly pre-profiled plate onto the cranium using self-tapping screws. The scalp tissue was then replaced and sutured. At post-operation recovery, results observed showed improved surgical results, reduced operation time and a reduce probability of complications.

7.9   JEWELRY INDUSTRY

The jewelry industry has traditionally been regarded as one which is heavily craft-based, and automation is generally restricted to the use of machines in the various individual stages of jewelry manufacturing. The use of RP technology in jewelry design and manufacture offers a significant breakthrough in this industry. In an experimental computer-aided jewelry design and manufacturing system jointly developed by Nanyang Technological University and Gintic Institute of Manufacturing Technology in Singapore, the SLA (from 3D Systems) was used successfully to create fine jewelry models [45]. These were used as master patterns to create the rubber molds for making wax patterns that were later used in investment casting of the precious metal end product (see Figure 7.16). In an experiment with the design of rings, the overall quality of the SLA models were found to be promising, especially in the generation of intricate details in the design. However, due to the nature of the step-wise building of the model, steps at the "gentler"

Figure 7.16: An investment cast silver alloy prototype of a broach (right), the full-scale wax pattern produced from the silicon rubber molding (center), and the two-time scaled SLA model to aid visualization (left)

slope of the model were visible. With the use of better resin and finer layer thickness, this problem was reduced but not fully eliminated. Further processing was found to be necessary, and abrasive jet deburring was identified to be most suitable [46].

Though post-processing of SLA models is necessary in the manufacture of jewelry, the ability to create models quickly (a few hours compared to days or even weeks, depending on the complexity of the design) and its suitability for use in the manufacturing process offer great promise in improving design and manufacture in the jewelry industry.

7.10   COIN INDUSTRY

Similar to the jewelry industry, the mint industry has traditionally been regarded as very labor-intensive and craft-based. It relies primarily on the skills of trained craftsmen in generating the "embossed" or relief designs on coins and other related products. In another experimental coin manufacturing system using CAD/CAM, CNC and RP technologies developed by Nanyang Technological University and Gintic Institute of Manufacturing Technology in Singapore, the SLA (from 3D Systems) was used successfully with a Relief Creation Software to create tools for coin manufacture [47]. In the system

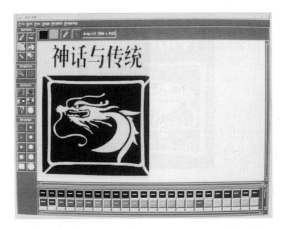

Figure 7.17: Two-dimensional artwork of a series of Chinese characters and a roaring dragon

involving RP technology, its working methodology consists of several steps.

Firstly, 2D artwork is read into ArtCAM, the CAD/CAM system used in the system, utilizing a Sharp JX A4 scanner. Figure 7.17 shows the 2D artwork of a series of Chinese characters and a roaring dragon. In the ArtCAM environment, the scanned image is reduced from a color image to a monochrome image with the fully automatic "Gray Scale" function. Alternatively, the number of colors in the image can be reduced using the "Reduce Color" function. A color palette is provided for color selection and the various areas of the images are colored, either using different sizes and types of brushes or the automatic flood fill function.

The second step is the generation of surfaces. The shape of a coin is generated to the required size in the CAD system for model building. A triangular mesh file is produced automatically from the 3D model. This is used as a base onto which the relief data is wrapped and later combined with the relief model to form the finished part.

The third step is the generation of the relief. In creating the 3D relief, each color in the image is assigned a shape profile. There are various fields that control the shape profile of the selected colored

Figure 7.18: Three-dimensional relief of artwork of the roaring dragon

region, namely, the overall general shape for the region, the curvatures of the profile (convex or concave), the maximum height, base height, angle and scale. The relief detail generated can be examined in a dynamic Graphic Window within the ArtCAM environment itself. Figure 7.18 illustrates the 3D relief of the roaring dragon artwork.

The fourth step is the wrapping of the 3D relief onto the coin surface. This is done by wrapping the three-dimensional relief onto the triangular mesh file generated from the coin surfaces. This is a true surface wrap and not a simple projection. The wrapped relief is also converted into triangular mesh files. The triangular mesh files can be used to produce a 3D model suitable for color shading and machining. The two sets of triangular mesh files, of the relief and the coin shape, are automatically combined. The resultant model file can be color-shaded and used by the SLA to build the prototype.

The fifth step is to convert the triangular mesh files into the STL file format. This is to be used for building the RP model. After the conversion, the STL file is sent to the SLA to create the 3D coin pattern which will be used for proofing of design [47].

7.11   TABLEWARE INDUSTRY

In another application to a traditional industry, the tableware industry, CAD and RP technologies are used in a integrated system to create better designs in a faster and more accurate manner. The general methodology used is similar to that used in the jewelry and coin industries. Additional computer tools with special programs developed to adapt decorative patterns to different variations of size and shape of tableware are needed for this particular industry [48]. Also a method for generating motifs along a circular arc is also developed to supplement the capability of such a system [49].

The general steps involved in the art to part process for the tableware include the following:

(1) Scanning of the 2D artwork.
(2) Generation of surfaces.
(3) Generation of 3D decoration reliefs.
(4) Wrapping of reliefs on surfaces.
(5) Converting triangular mesh files to STL file.
(6) Building of model by the RP system.

Two RP systems are selected for experimentation in the tableware system. One is 3D Systems' SLA, and the other is Helysis' LOM. The SLA has the advantages of being a pioneer and a proven technology with many excellent case studies available. It is also advantageous to use in tableware design as the material is translucent thus allowing designers to view the internal structure and details of tableware items like tea pots and gravy bowls. On the other hand, the use of LOM has its own distinct advantages. Its material cost is much lower and because it does not need support in its process (unlike the SLA), it saves a lot of time in both pre-processing (deciding where and what supports to use) and post-processing (removing the supports). Examples of dinner plates built using the systems are shown in Figure 7.19.

In an evaluation test of making the dinner plate prototype, it was found that the LOM prototype is able to recreate the floral details more accurately. The dimensional accuracy is slightly better in the LOM prototype. In terms of the build-time, including pre- and post-processing, the SLA is about 20% faster than the LOM process. However, with sanding and varnishing, the LOM prototype is found to be a better model which can be used later to create the plaster of Paris molds for the molding of the ceramic tableware (see Figure 7.20 for a tea-pot built using LOM). Apart from these technical issues, the

Figure 7.19: Dinner plate prototype built using SLA (left) and LOM (right)

Figure 7.20: LOM model of a tea pot (Courtesy of Champion Machine Tools, Singapore)

initial investment, operating and maintenance costs of the SLA are considerably higher than that of the LOM, estimated to be about 50% to 100% more.

In the ceramic tableware production process, the LOM model can be used directly as a master pattern to produce the block mold. The mold is made of plaster of Paris. The result of this trial is shown in Figure 7.21. The trials highlighted the fact that plaster of Paris is an extremely good material for detailed reproduction. Even slight imperfections, left

Figure 7.21: Block mold cast from the LOM model of the dinner plate
(Courtesy of Oriental Ceramics Sdn. Bhd., Malaysia)

after hand finishing the LOM model, are faithfully reproduced in the block mold and pieces cast from these molds.

Whichever RP technology is adopted, such a system saves time in designing and developing tableware, particularly in building a physical prototype. It can also improve designs by simply amending the CAD model and the overall system is easy and friendly to use.

REFERENCES

[1] Mueller, T., "Applications of stereolithography in injection molding," *Proceedings of the Second International Conference on Rapid Prototyping*, Dayton, USA, June 23–26, 1991, pp. 323–329.

[2] Wang, J.H., Lim, C.S., and Yeo, J.H., "CFD investigations of steady flow in Bi-leaflet heart valve," *Critical Reviews in Biomedical Engineering* **28**(1–2) (2000): 61–68.

[3] Chua, C.K., Hong, K.H., and Ho, S.L., "Rapid tooling technology (Part 1) — A comparative study," *International Journal of Advanced Manufacturing Technology* **15**(8) (1999): 604–608.

[4] Wilkening, C., "Fast production of rapid prototypes using direct laser sintering of metals and foundry sand," *Second National*

Conference on Rapid Prototyping and Tooling Research, UK, November 18–19, 1996, pp. 153–160.

[5] Tsang, H.B. and Bennett, G., "Rapid tooling — Direct use of SLA molds for investment casting," *First National Conference on Rapid Prototyping and Tooling Research*, UK, November 6–7, 1995, pp. 237–247.

[6] Atkinson, D., *Rapid Prototyping and Tooling: A Practical Guide*, Strategy Publications, UK, 1997.

[7] Chua, C.K., Hong, K.H., and Ho, S.L., "Rapid tooling technology (Part 2) — A case study using arc spray metal tooling," *International Journal of Advanced Manufacturing Technology* **15**(8) (1999): 609–614.

[8] Venus, A.D., Crommert, S.J., and Hagan, S.O., "The feasibility of silicon rubber as an injection mold tooling process using rapid prototyped pattern," *Second National Conference on Rapid Prototyping and Tooling Research*, UK, November 18–19, 1996, pp. 105–110.

[9] Shonan Design Co. Ltd., Temp-Less 3.4.3 — UV RTV process for quick development and fast to market, 1996.

[10] Schaer, L., "Spin casting fully functional metal and plastic parts from stereolithography models," *The Sixth International Conference on Rapid Prototyping*, Dayton, USA, June 1995, pp. 217–236.

[11] Male, J.C., Lewis, N.A.B., and Bennett, G.R., "The accuracy and surface roughness of wax investment casting patterns from resin and silicon rubber tooling using a stereolithography master," *Second National Conference on Rapid Prototyping and Tooling Research*, UK, November 18–19, 1996, pp. 43–52.

[12] Bettay, J.S. and Cobb, R.C., "A rapid ceramic tooling system for prototyping plastic injection moldings," *First National Conference on Rapid Prototyping and Tooling Research*, UK, November 6–7, 1995, pp. 201–210.

[13] Warner, M.C., "Rapid prototyping methods to manufacture functional metal and plastic parts," *Rapid Prototyping Systems: Fast Track to Product Realization*, 1993, pp. 137–144.

[14] Lim, C.S., Siaminwe, L., and Clegg, A.J., "Mechanical property enhancement in an investment cast aluminum alloy and metal-matrix composite," *9th World Conference on Investment Casting*, October 1996, San Francisco, USA, 17: 1–18.

[15] Clegg, A.J., *Precision Casting Processes*, Pergamon Press Plc., Oxford, England, 1991.

[16] Venus, A.D. and Crommert, S.J., "Direct SLS nylon injection," *Second National Conference on Rapid Prototyping and Tooling Research*, UK, November 18–19, 1996, pp. 111–118.

[17] Soar, R.C., Arthur, A., and Dickens, P.M., "Processing and application of rapid prototyped laminate production tooling," *Second National Conference on Rapid Prototyping and Tooling Research*, UK, November 18–19, 1996, pp. 65–76.

[18] Industrial Technology, "Laser sintering for rapid production," URL: http://www.industrialtechnology.co.uk/2001/may/eos.html.

[19] Wohlers Report 2000, *Rapid Prototyping and Tooling State of the Industry*, Wohlers Associates Inc., USA, 2000.

[20] Eyerer, P., "Rapid tooling — Manufacturing of technical prototypes and small series," *Mechanical Engineering* (1996): 45–47.

[21] Rapid Prototyping Report, *Sundstrand Aerospace Uses Laminated Object Manufacturing to Verify Large Complex Assembly* **5**(6), CAD/CAM Publishing Inc. (June 1995): 1–2.

[22] 3D Systems, "User Focus: AlliedSignal Aerospace, Stereolithography and QuickCast™ provide the winning combination for meeting critical deadline in AlliedSignal's development of the TFE 731-20 Turbo Fanjet Engine," 1993.

[23] Rapid Prototyping Report, *Sundstrand Power Systems Uses Selective Laser Sintering to Create Large Investment Casting Patterns* **5**(1), CAD/CAM Publishing Inc. (January 1995): 1–2.

[24] DTM Corporation Press Release, *SLS-Generated Polycarbonate Patterns Speed Casting of Intricate Aircraft Engine Part for Sundstrand Power Systems*, DTM Corporation, 1994.

[25] Rapid Prototyping Report, *Bell Helicopter Uses QuickCast to Fabricate Flight-Certified Production Casting* **5**(7), CAD/CAM Publishing Inc. (July 1995): 1–2.

[26] Rapid Prototyping Report, *Volkswagen Uses Laminated Object Manufacturing to Prototype Complex Gearbox Housing* **5**(2), CAD/CAM Publishing Inc. (February 1995): 1–2.

[27] 3D Systems, *Stereolithography Provides a Symphony of Benefits, The Edge* **6**(2) (1995): 6–7.

[28] 3D Systems, The Winner: Mercedes-Benz AG, 1994 European Stereolithography Excellence Awards, 1995.

[29] Rapid Prototyping Report, *Ford Uses Stereolithography to Cast Production Tooling* **4**(7), CAD/CAM Publishing Inc. (July 1995): 1–3.

[30] 3D Systems, The Winner: Ford Motor Company, 1994 Stereolithography Excellence Awards, 1995.

[31] Mahoney, D.P., "Rapid prototyping in medicine," *Computer Graphics World* **18**(2) (1995): 42–48.

[32] Swaelens, B. and Kruth, J.P., "Medical applications in rapid prototyping techniques," *Proceedings of the Fourth International Conference on Rapid Prototyping*, June 14–17, 1993, pp. 107–120.

[33] Jacobs, A., Hammer, B., Niegel, G., Lambrecht, T., Schiel, H., Hunziker, M., and Steinbrich, W., "First experience in the use of stereolithography in medicine," *Proceedings of the Fourth International Conference on Rapid Prototyping*, June 14–17, 1993, pp. 121–134.

[34] Adachi, J., Hara, T., Kusu, N., and Chiyokura, H., "Surgical simulation using rapid prototyping," *Proceedings of the Fourth International Conference on Rapid Prototyping*, June 14–17, 1993, pp. 135–142.

[35] Koyayashi, M., Fujino, T., Chiyokura, H., and Kurihara, T., "Preoperative preparation of a hydroxyapatite prosthesis for bone defects using a laser-curable resin model," *The Inaugural Congress of the International Society for Simulation Surgery*, 1992.

[36] Kaneko, T., Kobayashi, M., Tsuchiya, Y., Fujino, T., Itoh, M., Inomata, M., Uesugi, M., Kawashima, K., Tanijiri, T., and Hasegawa, N., "Free surface three-dimensional shape measurement system and its application to Mictotia ear reconstruction," *The Inaugural Congress of the International Society for Simulation Surgery*, 1992.

[37] Chow, K.Y., "Development of a direct link between a laser digitizer and a rapid prototyping system," Final year thesis, Nanyang Technological University, Singapore, 1996.

[38] Vacanti, C.A., Kim, W., Upton, J., Vacanti, M.P., Mooney, D., Schloo, B., and Vancanti, J.P., "Tissue-engineered growth of bone and cartilage," *Transplantation Proceedings* **25**(1) (1993): 1019–1021.

[39] Freed, L.E., Marquis, J.C., Nohria, A., Emmanuel, J., Mikos, A.G., and Langer, R., "Neo-cartilage formation in in-vitro and in-vivo using cells cultured on synthetic biodegradable polymers," *Journal of Biomedical Materials Research* **27**(1993): 11–23.

[40] Yang, S.F., Leong, K.F., Du, Z.H., and Chua, C.K., "The design of scaffolds for use in tissue engineering (Part 1)," *Traditional Factors Tissue Engineering* **7**(6) (2001).

[41] Yang, S.F., Leong, K.F., Du, Z.H., and Chua, C.K., "The design of scaffolds for use in tissue engineering (Part 2)," *Rapid Prototyping Techniques Tissue Engineering* **8**(1) (2002).

[42] Lim, C.S., Lin, S.C., Eng, P., and Chua, C.K., "Rapid prototyping and tooling of custom-made tracheobronchial stents," *International Journal of Advanced Manufacturing Technology* **20**(1) (2002): 44–49.

[43] Lim, C.S., Chandrasekeran, M., and Tan, Y.K., "Rapid tooling of powdered metal parts," *International Journal of Powder Metallurgy* **37**(2) (2001): 63–66.

[44] Lim, C.S. and Chandrasekeran, M., "A process to rapid tool porous metal implants," Internal Report, Nanyang Technological University, Singapore (February 1999): 1–6.

[45] Lee, H.B., Ko, M.S.H., Gay, R.K.L., Leong, K.F., and Chua, C.K., "Using computer-based tools and technology to improve jewelry design and manufacturing," *International Journal of Computer Applications in Technology* **5**(1) (1992): 72–88.

[46] Leong, K.F., Chua, C.K., and Lee, H.B., "Finishing techniques for jewelry models built using the stereolithography apparatus," *Journal of the Institution of Engineers* **34**(4) (1994): 54–59.

[47] Chua, C.K., Gay, R.K.L., Cheong, S.K.F., Chong, L.L., and Lee, H.B., "Coin manufacturing using CAD/CAM, CNC and rapid prototyping technologies," *International Journal of Computer Applications in Technology* **8**(5/6) (1995): 344–354.

[48] Chua, C.K., Hoheisel, W., Keller, G., and Werling, E., "Adapting decorative patterns for ceramic tableware," *Computing and Control Engineering Journal* **4**(5) (1993): 209–217.

[49] Chua, C.K., Gay, R.K.L., and Hoheisel, W., "A method of generating motifs aligned along a circular arc," *Computers & Graphics: An International Journal of Systems and Applications in Computer Graphics* **18**(3) (1994): 353–362.

PROBLEMS

1. How is application of RP models related to the purpose of prototyping? How does it also relate to the materials used for prototyping?

2. List the types of industries that RP can be used in. List specific industrial applications.

3. What are the finishing processes that are used for RP models and explain why they are necessary?

4. What are the typical RP applications in design? Briefly describe each of these applications and illustrate them with examples.

5. What are the typical RP applications in engineering and analysis? Briefly describe each of them and illustrate them with examples.

6. Describe how RP models can be used for pre-surgical operation planning. Use appropriate examples to illustrate your answer.

7. Why and in what circumstances would RP be considered to assist implant fabrication?

8. Describe two examples of how rapid prototyping and tooling techniques would be preferred over conventional methods in the improvement of patient care.

9. How would you differentiate between the following types of rapid tooling processes: (a) direct soft tooling, (b) indirect soft tooling, (c) direct hard tooling, and (d) indirect hard tooling.

10. Explain how a RP pattern can be used for vacuum casting with silicon molding. Use appropriate examples to illustrate your answer.

11. What are the ways the RP pattern can be used to create the injection mold for plastic parts. Briefly describe the processes.

12. Compare and contrast the use of RP patterns for
    (i)   casting of die inserts,
    (ii) sand casting, and
    (iii) investment casting.

13. What are the RP systems that are suitable for sand casting? Briefly explain why and how they are suitable for sand casting?

14. Compare the relative merits of using LOM parts with SLA parts for investment casting.

15. Explain whether RP technology is more suitable for "high technology" industries like aerospace than it is for consumer

products industries like electronic appliances. Give examples to substantiate your answer.

16. Explain how RP systems can be applied to traditional industries like the jewelry, coin and tableware industries.

Chapter 8
EVALUATION AND BENCHMARKING

8.1 USING BUREAU SERVICES

The best way to try out RP is with a service bureau that owns and operates one or several RP systems. RP equipment is fairly expensive (typically upwards of US$50 000) and the cost of operator training, materials and installation of the equipment can easily double this cost. Depending on the volume of prototyping work in your company, it will probably not be enough to justify the acquisition of such a system. Thus, it may be more economical to engage a service bureau.

When choosing a service bureau, it is worthwhile to consider the following points:

(1) Type of material needed for the prototype,
(2) Size of the prototype,
(3) Accuracy,
(4) Surface finish,
(5) Experience of service bureau,
(6) Location of service bureau for communication and coordination of jobs,
(7) Provision of secondary processes such as machining or sandblasting,
(8) Cost.

Typically, a service bureau bases its charges on the following items:

(1) Modeling and preparation of model for building (if necessary),
(2) Execution of program to convert computer model into STL file,
(3) "Slicing" time in converting STL file into cross sectional data,
(4) The amount of machine time required to make the part,
(5) The setup time in preparing the apparatus,
(6) The actual amount of material used to make the part,

(7) Post fabrication assembly such as gluing together several smaller pieces,

(8) Secondary processes such as machining or sand blasting.

Needless to say, the company can save cost on the first two items if it has its own CAD facility and CAD-RP interface. Fees from service bureaus can range from a few hundred dollars for a simple part to thousands of dollars for large and complex prototypes. The Wohler's Report [1], published annually, lists service bureaus worldwide.

8.2   SETTING UP A SERVICE BUREAU

There are three major considerations in assessing the economic feasibility of setting up the proposed RP service bureau. These are as follows:

(1) Is there sufficient demand for rapid prototypes in the country (or region) to justify the establishment of an additional service bureau?

(2) Can the proposed facility operate at a profit with the current market price for RP parts, proposed level of output and the investment necessary to establish the proposed facility?

(3) Will the profit realized from the operation of the proposed facility justify the investment?

To illustrate how a study on the setting up of a service bureau can be carried out, a case will be presented. The figures quoted in the case study are fictitious due to the confidentiality of such an actual study. However, they are sufficient to illustrate the workings of such a study.

8.2.1   Preliminary Assumptions

Several basic assumptions are required in developing the substantive materials, which will be analyzed to reach the conclusions. First, the size of operation must be assumed. If it is assumed that the initial size of the operation will be small, this could mean that only one RP system could be considered for acquisition. Accordingly, the skilled manpower to be employed (or re-deployed) to provide the service, both technical and administrative, will only be sufficient for that operation.

Secondly, it is assumed that the proposed service bureau would concentrate on the industries represented in the country (or region). The market potential existing in the country or region and the anticipated size of the operation warrant such an assumption.

The third assumption deals with the exact nature of the operation. For example, if it is assumed that the service bureau would not be an integrated operation, then there will be no additional processing of parts such as secondary milling, grinding processes, etc. Thus, these accessories and equipment need not be computed in the study.

A final assumption is that the service bureau would operate at a predetermined level of output. Certain basic facilities are required to start such an operation. However, the utilization of such equipment may vary, though not considerably, due to seasonal variations in the availability of big projects from major clients. This specified level of operation would entail an average-sized project (one or more prototypes, may be identical or different part) lasting for three working days. Taking 262 working days per year, about 87 projects can be undertaken annually. For the purpose of calculation, it is assumed that each project handles two parts as all RP systems can build numerous parts simultaneously. This production (or sales) level is vital to the analysis that follows. Operating at some other level of output than the one specified would substantially influence cost and revenues and would not be reflected in the analysis of the study.

It should be pointed out that no attempt is made to analyze the managerial abilities of administrative personnel of the service bureau. The profitability of any business operation is however dependent on the possession of adequate managerial abilities by those personnel responsible for the decision making within the organization.

8.2.2   Market Potential for Prototyping

8.2.2.1   *Consumer Demand*

The demand of prototypes in the country may or may not be available directly. Sometimes, though no official figures of the actual demand of prototypes are available, other forms of evidence can point or indicate the state of demand, such as:

(1) Many multi-national or foreign companies re-locating their design activities to the country.
(2) Shorter product life cycle for many products (e.g., cellular phone, pager, etc.).
(3) More *designed-in-country* products.
(4) Government's or local agencies' substantial increase in R&D expenditure so as to promote research, which in turn will increase the demand for the services of prototypes.

A good source of statistics is from census or trade reports. For instance, in Singapore, from the *Report on Census of Industrial Production, 1993* [2], it is possible to establish figures relating to the sales of various products and services.

From numerous RP reports, case studies and proceedings published, it can be inferred that many products and services can use the service bureau. These products and services are shown in Table 8.1, along with the number of establishments and total sales figures.

However, the sales figure is a total figure and will not reflect accurately the same figure that can be replaced by RP. Nevertheless, the figures provide a useful gauge of the total number of establishments and sales volume.

To be a step closer to realizing the dollar value of prototype demand, a set of data is collected from the industries most active in RP. The collection of data is carried out while maintaining the confidentiality of the respondents. Table 8.2 shows a breakdown of demand by major industry group, which are likely to be most active in RP. An average budget of US$1000 is assumed for the building of a RP part. Therefore, column 5 is obtained by the multiplication of columns 2, 3 and 4 and US$1000. This estimate of US$1000 is an average as it is reasonable to assume that some more complex and bigger parts will cost more. On the other hand, simpler parts will cost slightly lesser.

Table 8.1: Sales volume of product and services [2]

Major Industry Group	Establishments	Sales (US$ '000)	Application Area of RP
Dies, molds, tools, jigs and fixtures	130	299 274	Dies, molds and tools
Connectors	10	279 049	Product design and functional testing
Electrical household appliances	7	517 029	Product design and fitting
Electronic products and components	247	32 378 744	Product design and mock-up
Computer peripheral equipment	20	3 590 355	Product design and functional testing
Communication equipment	9	1 422 222	Product design and fitting
Television sets and sub-assemblies	5	1 587 638	Product design and fitting
Audio and video combination equipment	11	3 089 532	Product design and fitting
Motor vehicle parts and accessories	14	84 491	Product design and functional testing
Surgical and medical instruments	12	364 482	Product design
Watches and clocks	8	191 324	Product design
Jewelry	57	338 653	Product design and tooling
Toys	13	218 932	Product design and tooling
Perfumes, cosmetics and toilet preparations	11	31 778	Product design
Plastic household and kitchen ware	7	11 139	Product design and fitting
Plastic bottles, boxes and containers	29	145 822	Product design
Plastic precision parts	116	815 743	Product design and fitting
Pottery, earthenware and glass products	12	173 617	Product design and tooling
Footwear	27	47 632	Product design and tooling

Table 8.2: Estimated demand for RP in dollar value

Major Industry Group	Establishments	Number of Models per Year	Number of Parts per Model	Sales (US$ '000)
Connectors	10	6	2	120
Computer peripheral equipment	20	8	10	1600
Electrical household appliances	7	12	8	672
Communication equipment	9	16	4	576
Television sets and sub-assemblies	5	6	18	540
Audio and video combination equipment	11	8	12	1056
Motor vehicle parts and accessories	14	2	36	1008
Total				5572

The above estimated demand for RP is 5572 parts and US$5 572 000 in dollar value. Suppose there are six service bureaus in the country producing at the rate of two RP parts per week, the total output of present service bureaus is:

Total output of present service bureaus
= 6 service bureaus × 2 RP parts/week/service bureau
  × 52 weeks/year
= 624 parts per year.

This number is far from meeting the RP demands of the industries. Therefore, the analysis of the prototyping business in the country compared to the total output of the present service bureaus in the country shows a net excess of consumption when stated in prototype number (5572 − 624) = 4948 parts. In terms of equivalence in dollars, this net excess of demand is just under US$5.0 million. This figure represents the potential for the service bureau.

The growth in consumer demand for rapid prototypes is extremely promising based on a study by Terry Wohlers [3]. Revenues collected from both sales and service bureau have increased over the years. However, the increase in revenues from service bureau is more spectacular. Revenues jumped exponentially since 1993 and continued to repeat similar performances in subsequent years.

Such a phenomenon can only be explained by two reasons. First, there is more industrial awareness now than before. The number of annual conferences devoted to Rapid Prototyping in USA and Europe, which started since 1990, has grown significantly. Many more technical papers, journal and other articles have been written on RP.

Secondly, the capabilities of RP systems have improved tremendously since its early beginning in 1988 — in terms of material range, accuracy, mechanical properties, industrial applications, etc. These improvements have won over many skeptics.

While there is no formal study done which indicates that the trend experienced in the United States is being duplicated worldwide, it is believed that other regions will experience more or less a similar type of growth.

8.2.2.2 *Application Areas of RP Systems*

This is important because there may be a strong correlation between the types of RP systems and the industry in which the prototypes are used. For instance, the connector industries use prototypes which must meet certain requirements (testability, form and fit, etc.) and these requirements may be fulfilled easily by, say, RP system A and not so by RP system B. It should, however, be noted that with the trend in RP systems' advancements, this gap is rapidly narrowing.

8.2.2.3 *Competitors*

The competitors in the country or region are those which currently operate a service bureau. The six service bureaus are shown in

Table 8.3: RP service bureau

RP systems used in service bureau	3D Systems' SLA	CMET's SOUP	Sony's SCS	Cubic's LOM	Stratasys' FDM	DTM's SLS and 3D Systems' SLA
Name of service bureau	AAA	BBB	CCC	DDD	EEE	FFF
Service available since	1990	1994	1992	1993	1995	1995

Table 8.3. Note that the table is tabulated column-wise by the type of RP systems used in servicing clients. This arrangement is important as it will be shown later that the type of RP system is pivotal to the area of applications and indirectly responsible for the potential number of clients whom the service bureau can serve.

In their years (or months) of operation, it is not possible to obtain their outputs as these are confidential information. However, one can safely estimate that the current usage exceeds an average order of two parts per week.

An analysis of the level of expertise and experience and the number of operational staff available in the service can also be carried out. This relates directly to their capabilities and limitations in handling projects. Table 8.4 shows the number of people involved, the level of experience and the clients served by the service bureau.

8.2.2.4 *Competitive Analysis*

The strengths and weaknesses of the competitors are next analyzed. Considerations include the nature of the service bureau such as whether it is primarily a vendor or distributor, or an academic or research entity. In such instances, they may be weak in providing a total integrated solution. The priority of a vendor/distributor is focused on sales

Table 8.4: Service bureau staff, experience and clients

Service Bureau	RP Systems	Number of Staff	Clients	Experience
AAA	SLA	Five engineering staff	Jewelry, souvenir, medical, consumer products, audio, video, computer and telecommunications industry	Most experienced of all in RP; goal is to transfer technology
BBB	SOUP	Not known (20 persons in company)	Telecommunications, computer, audio, video, industrial and consumer products industry	Newcomer in RP, however has been in tool and mold business
CCC	SCS	Two engineering staff	Consumer electronics, consumer products, automobile accessories, audio and video industry	Second-most experienced; meets own needs; however, also its main disadvantage
DDD	LOM	Three engineering staff	Telecommunications, computer, audio, video, industrial and consumer products industry	Not main business
EEE	FDM	Not known (Six persons in company)	Telecommunications, computer, audio, video, industrial and consumer products industry	Machine not delivered yet; however has been in mold and electrode making business
FFF	SLS	Not started yet	NA	Machine not delivered yet

whereas; the academic or research entity's primary concern is education, teaching and research. Another consideration is the availability of additional facilities and equipment in supporting basic engineering works, which are important in providing a holistic approach to engineering solutions.

8.2.2.5   *Marketing Strategies*

Potential clients do not know who you are and what you can offer when your services are first launched. Promotional activities must accomplish this for you. There are two basic forms of promotional activities:

(1) *Publications*. One of the most effective ways is to disseminate information through publications. There are five types of publications and they are described as follows:

   (a) *Service Bureau Brochure*. First, a Service Bureau Brochure will have to be designed and created. No efforts should be spared in making it top-class-colorful, gloss-finished, and graphics illustrated. Important details like total engineering solutions, lead-time and one-stop service should be emphasized. The brochure should not be clouded with technical details, but should be designed professionally and be more business-oriented. A single folded sheet brochure is preferred.

   (b) *Technical Brief*. This is where the technical details — process description, material properties and composition, etc. — can be included. This brief information is especially useful for the technically inclined clients. However, this brief information should preferably be a single sheet. Alternatively, the Service Bureau Brochure and Technical Brief can be combined into an A3-sized paper folded in the middle.

   (c) *Technical Article*. In times where participation at a conference, seminar or workshop requires a submission of an article on the service or its experience on special application projects, this will come in useful. The length of the article can be about six to ten pages.

   (d) *Technical Video and/or Disk (or CD)*. Most RP systems come with a video showing technical details and possibly case studies and interviews with clients. It can be used directly for your purpose. Alternatively, a CD using multi-media technology can be created. This, however, will have to be carried out by the company, as many RP systems do not come with this.

   (e) *Internet Website with Easily Identifiable Uniform Resource Locator (URL)*. Many potential clients will have access to the Internet, and the information of service bureau must be made available readily. An easily identifiable URL or website address will be necessary. Professional assistance in setting up the website may be necessary. Complete information must be made available, including details like total engineering solutions, lead-time and one-stop services which must be emphasized. Ease of navigation through the pages will be essential. Multi-media design created on the video CDs can also be edited and included here.

(2) *Publicity Activities*. Publicity activities are useful in that you can publicize through an attentive audience. It may be organized solely by your company or in conjunction with interested parties. Where it is organized by some other organizations, it is worthwhile to keep in contact with the relevant active organizations so that you are kept informed and where possible, be invited for participation. There are three types of publicity activities:

(a) *Technical Talk*. This is a very common activity organized by many engineering societies. However, some societies discourage sales talk and therefore, the subject contents will have to be technical. In many instances, the talk is free for members and thus, the company may have to sponsor the event. This event can be held in tertiary institutions or at hotels, depending on your budget. It is typically one to one-and-a-half hours' in the evening, and should include refreshments.

(b) *Exhibition*. Participation in relevant, major exhibitions held nationally or regionally will be useful too. It is cautioned that substantial costs will be incurred (mainly transportation costs and rental of exhibition booth). An evaluation should be carried out to decide if it is worthwhile. Exhibitions are typically several days' affair and would mean disruption of work for at least two staff to man the booth.

(c) *Seminars/Conferences*. Like a technical talk, speaking at a conference or seminar requires a fair amount of experience and knowledge of your product. Unlike the technical talk, submission of an article is required. Typically, a notification of such an event comes through a "Call for Papers," followed by submission of an abstract, and if accepted, the full manuscript. Every two years, there are several relevant conferences devoted to "Manufacturing", "Automation" and "Computer Integrated Manufacturing (CIM)". Since 1990, many dedicated international conferences and seminars on "Rapid Prototyping" have also been held.

8.2.2.6   *Competency and Preparation*

Potential clients do not know how good you are and therefore whatever contact you make with them must demonstrate your competency and reliability. Some of the possible contacts with potential clients may be through phone calls, visits, demonstrations, benchmarking tests, etc. Preparatory activities include really knowing your products, putting up big posters (create your own) describing the process, using impressive show cases to display industrial parts made by the RP systems, as well as other secondary value-added processes put in by your company.

8.2.3   Cost of the Service Bureau Setup

8.2.3.1   *Land and Building Costs*

Land and building costs are important in any operation. Very likely, the land and building are to be leased. Location will be one very significant consideration as it should be as close to the client as possible. In this case study, the lease on land and building of an appropriate size is taken to be US$20 000 per annum.

8.2.3.2   *Main Equipment Costs*

The list and current prices of equipment necessary to outfit the proposed facility is shown in Table 8.5. Total equipment costs include machine, materials, accessories and post cure equipment (if applicable). Typically, pricing also includes installation, commissioning and training.

Table 8.5: Cost of RP equipment (*1995 prices)

RP Systems	Estimated Price*
SLA	US$300 000
SGC	US$295 000
SLS	US$366 000
LOM	US$90 000
FDM	US$120 000
SOUP	US$600 000
SCS	US$380 000

8.2.3.3 *Optional Equipment Costs*

The optional equipment include a workstation and a CAD/CAM package such as Pro-Engineer. Table 8.6 shows an example of the likely cost of such equipment. Total optional equipment cost with an IBM-compatible personal computer is US$33 000.

Table 8.6: Cost of optional equipment
(*one set of software and hardware, at 1995 prices)

CAD/CAM	Estimated Price*
Software	Pro-Engineer Basic software: US$25 000
Training	Basic: US$4000 (5 days) Surface: US$2000 (3 days)
Hardware	Pentium: US$2000 or SGI: US$30 000

8.2.4 Comparison and Selection of RP Systems

A point-by-point comparison of the various RP systems under consideration is carried out. It is worthwhile to note that there are strengths and weaknesses for every system and the weaknesses of the systems are constantly improved upon by the vendors.

From the analysis of various factors, suppose that LOM is recommended for the service bureau for the following supporting reasons:

(1) The machine cost is amongst the lowest amongst all its competitors,
(2) Its applications are substantially wide, based on sales figures,
(3) It has also the lowest material costs,
(4) The process is clean and can operate in an office environment,
(5) Material need not be shielded from sunlight and does not have a foul smell,
(6) Supports are not required in the process.

8.2.5   Capital Investment Requirements

The proposed venture can now be stated in terms of its initial investment requirements. The initial capital required will include the RP machine and the CAD/CAM system, inclusive of all training needs.

(a) LOM   : US$140 000
(b) CAD/CAM : US$33 000
     Total   : US$173 000

8.2.6   Revenues, Expenses and Return on Investment

8.2.6.1   *Revenues*

The revenue derived from operations will come from the revenues derived from the production of prototypes. In the future plans for the service bureau, other sources of revenue can come from the production of molds, tools and dies from the master pattern of the prototypes. This will complement the company's mission to provide a total solution package to clients in meeting a diverse range of manufacturing applications.

The level of output has already been pre-determined to be 87 projects per annum. This project is assumed to be of average size and will take three days to complete. It is also assumed that the project will involve two prototypes to be produced. The stages of production will include model checking, model building and model post processing. Other activities such as CAD modeling, file transfer and secondary processes will be excluded. From industrial data taken from competitors, it is fair to assume that an average project will cost US$1000. Therefore, annual revenue will amount to 87 × 2 × US$1000 = $174 000.

8.2.6.2 *Operating Expenses*

Three major categories of expenses are discussed in this section: cost of expendable, salaries, and depreciation and maintenance.

(1) *Cost of Expendable.* One of the major operating costs is the cost of raw materials. Raw materials may be liquid resin for the case of SLA, a roll of paper for LOM or a roll of filament for FDM. The price for each case varies. The recommended LOM uses paper, the cost of paper is US$100 per roll. This roll has a width of 210 mm and length of 150 m. For big parts, this roll of paper can be used to build 2 parts. It can build three to four parts for smaller components. Assume, therefore that each roll of paper can be used for every 2 projects. For 87 projects in a year, paper cost per annum works out to be (US$100/roll × 87 projects/2 projects/roll = US$4350).

 The other major expendable is the laser, if used. The laser is used in RP systems such as SLA, LOM and SLS. The price for each system varies. As before, the laser used in LOM is used for the calculations. Assume that the laser costs US$20 000 and can last 10 years. Laser cost per annum works out to be (US$20 000/ 10 years = US$2000). Other costs, if any, may be assumed to be negligible. Therefore, the summary of expendable per year = US$4350 + US$2000 = US$6350.

(2) *Salaries.* Assume that the number of personnel required for the operation is two. Administrative and marketing functions are presumed to be already in existence. The functions of these two personnel are predominantly technical and both are equally proficient in the entire process of RP so that each can cover the other's duty without any problem or to a cause in delay in the project.

However, for proper management and control, one should be an engineer and the other an engineering assistant. In this way, the engineer can focus on scheduling and prioritizing projects and the assistant on the project proper. The following figures show the yearly salaries:

(i) Engineer   : US$20 000
(ii) Engineering Assistant : US$12 000
      Total   : US$32 000

Total annual salaries amount to US$32 000. Other employee expenses are also included in the above salaries. These could be insurance, hospitalization and other fringe benefits and the total wage per annum is US$35 000.

(3) *Depreciation and Maintenance.* Assume that several other categories of expenses (such as utilities, accounting charges, bad debt expenses) amount to 10% of depreciation of the equipment. Total annual capital expense is therefore depreciate to 110%. If the equipment is be depreciated over a 15-year period, the annual depreciation expense would total US$173 000/15 years = US$11 533. The total annual capital expense is US$11 533 × 1.1 = US$12 686. The annual maintenance of the LOM and Pro-Engineer software is 15%, i.e., 0.15 × US$25 000 = US$3750.

8.2.6.3 *Pro Forma Income Statement*

The results of the analysis of anticipated revenues and expenses are next used to develop the pro forma income statement shown in Table 8.7. Expenses total US$77 786. This amount is subtracted from the total revenues to determine profit before taxes. Company income taxes are then subtracted to determine the net profit after taxes, which is US$70 236.

8.2.6.4 *Return on Investment Analysis*

The rate of return on investment is computed by dividing net profit after taxes of US$70 236 by the total investment (including

Table 8.7: Pro forma income statement

Sales (total revenues)		US$174 000
Lease	(US$20 000)	
Cost of expendable	(US$6350)	
Salaries	(US$35 000)	
Depreciation	(US$12 686)	
Maintenance	(US$3750)	
Total expenses		(US$77 786)
Profit before income taxes		US$96 214
Corporate taxes (27%)		US$25 978
Net profit after taxes		US$70 236

capital investment and annual operating expense less depreciation) of US$38 100. The rate of return = (US$70 236/US$238 100) × 100% = 29.5%.

Assuming a 15% minimum acceptable rate of return and since the calculated value of the rate of return of 29.5% is greater than 15%, the proposed venture is deemed economically viable.

8.3   TECHNICAL EVALUATION THROUGH BENCHMARKING

The execution of the benchmark test is a traditional practice, and necessary for all kinds of highly productive and expensive equipment such as the CAD/CAM workstation, CNC machining center, etc. Wherever a relatively broad spread of possibilities is offered to meet specific users' requirements, the execution of benchmark test is absolutely necessary. The dynamic development and increased range of commercially available RP systems on offer (currently more than 20 different types of equipment worldwide, partly in different types of equipment, partly in different sizes) mean objective decision making is essential. In analyzing the benchmark test piece, some tests have to be conducted including visual inspection and dimension measurement.

8.3.1   Reasons for Benchmarking

Generally, benchmarking serves the following purposes:

(1) It is a valuable tool for evaluating strengths and weaknesses of the systems tested. Vendors have to produce the benchmark models in response to requests from potential buyers. In doing so, they will not have the choice to demonstrate what they want, but have to show what are requested. Consequently, the vendors cannot hide the limitations of the system. It is also a rigorous and therefore more revealing means of testing so that the potential buyer can verify the claims of the vendor.

(2) Since the benchmark model is specifically designed by the potential buyer, it can be custom-made to its own requirements and needs. For example, in the case of a company that makes parts, which frequently have very thin walls, then the ability of the system to produce accurately built thin walls can be tested, measured and verified.

(3) Benchmarking has also become a means for helping various departments within a company to comprehend what the RP system can do for them. This is vitally important in the context of concurrent engineering whereby designers, analysts, and manufacturing engineers work on the product concurrently. Today, a RP system's application areas extend beyond design models, to functional models and manufacturing models.

(4) Sometimes, a benchmark test may also help to identify applications for a RP system, which had not previously been considered. Though this is not really a primary motivational factor for benchmarking, it is nevertheless a side benefit.

8.3.2   Benchmarking Methodology

There are four steps in the proposed benchmarking methodology:

(1) Deciding on the benchmarking model type.
(2) Deciding on the measurement to be made.

(3) Recording time and measurements, tabulating and plotting the results.
(4) Analyzing and comparing results.

8.3.2.1 *Deciding on the Benchmarking Model Type*

In general, rapid prototyping benchmarking models can be categorized according to the following types:

(1) *Typical Company's Products.* This is probably the most common type since the company needs to confirm how well its products can be prototyped and whether its requirements can be fulfilled. The company is also in the best position to comment on the results since it has intimate product knowledge. Examples are turbine blades, jewelry, cellular phones, etc.
(2) *Part Classification.* According to M.B. Wall [4], a classification scheme based on general part structures is applicable for rapid prototyping parts. This classification scheme based on part structure has ten part classes as seen in Table 8.8.
   Figure 8.1 shows some examples of such a classification scheme. The part sizes and part structures are most related to shrinkage, distortion and curling effects.

Table 8.8: Part structure classification scheme of ten part classes

Part Class Number	Part Structure
Part Class 1	Compact parts
Part Class 2	Hollow parts
Part Class 3	Box type
Part Class 4	Tubes
Part Class 5	Blades
Part Class 6	Ribs, profiles
Part Class 7	Cover type
Part Class 8	Flat parts
Part Class 9	Irregular parts
Part Class 10	Mechanisms

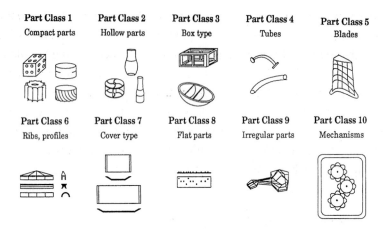

Figure 8.1: Part structure classification scheme for RP systems

(3) *Complex and Hybrid Parts.* Alternatively, determined complex parts can be designed, aiming to test the performance of available systems in specific aspects. Furthermore, this category can include a hybrid combining types 1 and 2 above.

8.3.2.2 *Deciding on the Measurements*

In deciding on the measurements, it must be stressed that, as far as possible, the benchmark model should:

(1) be relatively simple and designed with low expenditure,
(2) not utilize too much material, and
(3) allow simple measuring devices to determine the measurements.

In general, two types of measurements can be taken namely, main (large) measurements and detailed (small) measurements.

8.3.2.3 *Recording Time and Measurements, Tabulating and Plotting the Results*

The measurement results are based on the deviations of the built part from the CAD model. These deviations of both the main and detailed measurements are tabulated for each of the RP systems. Subsequently, the results can be plotted to show graphically the performance of the systems in a single diagram. Thus, for the main measurements, one can visually compare the systems' performance from the main measurement diagram and similarly for the detailed measurement diagram.

The time results are based on three components — data preparation, building time and post-processing. The total time is based on the sum of these three components. A table of all four time results can be tabulated with each RP system alongside another.

8.3.2.4 *Analyzing and Comparing Results*

In a general analysis, the deviations of all systems may or may not be in an acceptable range. The evaluator of the systems usually decides this acceptable range. The time component by itself gives one an idea of the length required for a task and directly affects the cost factor. Therefore, the time data can become useful for a full economic justification and cost analysis.

In comparison, one can determine, based on the time and measurement results, the strengths and weaknesses of the systems. In arriving at a conclusion, the evaluator must only consider the benchmarking results as *a component* of the overall evaluation study. The benchmarking results should never be taken above as the only deciding factor of an evaluation study.

Finally, the above approach ignores the human aspects. For example, the skills, expertise and experience of the vendors' operators are not accounted for in the benchmark.

8.3.3 Case Study

8.3.3.1 *The Button Tree Display*

With the compliments of Thomson Multimedia in Singapore, the results of a benchmarking study involving five machines are made available here. The test piece is a button tree display that is mounted in between the front cabinet of a hi-fi set and printed circuit board (PCB) as shown in Figure 8.2. The button tree display has a frame of length 128 mm and width 27 mm. It consists of three round buttons joined to the frame by 0.6 mm hinges. The buttons have "legs" that contact the tact switch on the PCB when it is depressed. There are also locating pins for location and the light emitting diode (LED) holder on the frame. Catches are made from the side of the frame so that the button tree display can hold down firmly to the PCB. The five different test pieces are made from the principals of SGC, LOM, SOUP, SCS and FDM. A photograph of each of the test pieces is shown in Figures 8.3, 8.4, 8.5, 8.6 and 8.7, respectively.

The measurements taken are linear, radial and angular dimensions. Coordinate measuring machine (CMM), profile projector and vernier caliper are used for the measurements. When choosing the type of dimensions to measure, a few criteria are considered:

(1) the overall dimensions are to be included,
(2) the important dimensions that will affect the operation of the button tree,

Figure 8.2: Button tree display is mounted on the front cabinet of a hi-fi

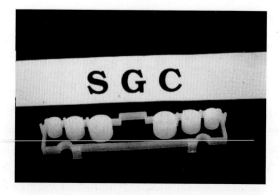

Figure 8.3: Benchmark test piece made from Solid Ground Curing (SGC)

Figure 8.4: Benchmark test piece made from Laminated Object Manufacturing (LOM)

Figure 8.5: Benchmark test piece made from Solid Object Ultraviolet Plotter (SOUP)

Figure 8.6: Benchmark test piece made from Solid Creation Systems (SCS)

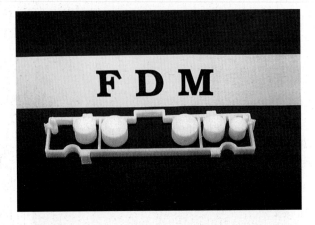

Figure 8.7: Benchmark test piece made from Fused Deposition Modeling (FDM)

(3) there must be a variety of dimensions,

(4) there must be sufficient main and detailed dimensions to plot the graph, and give an indication of which method is superior.

Figures 8.8, 8.9 and 8.10 show the technical drawings for the button tree display. The dimensions of the test piece are divided into three parts. The first part included dimensions are taken from the frame,

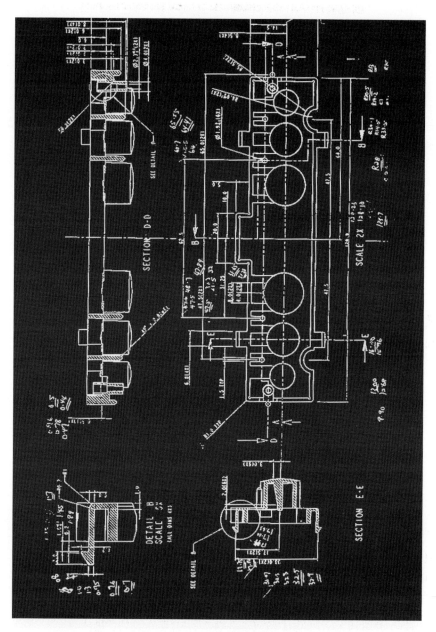

Figure 8.8: Working drawing showing the front and plan views of the
Button Tree Display

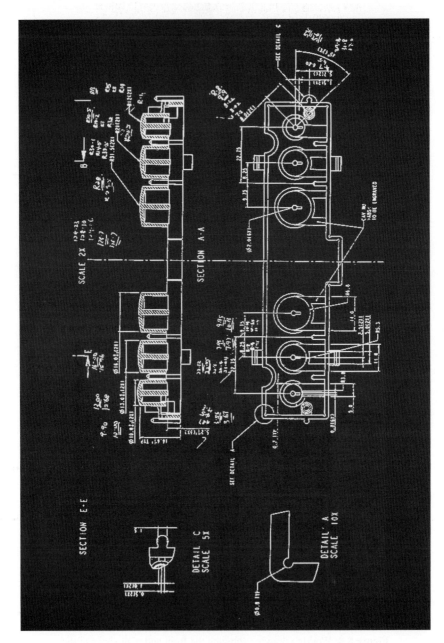

Figure 8.9: Working drawing showing the sectional view of the Button Tree Display

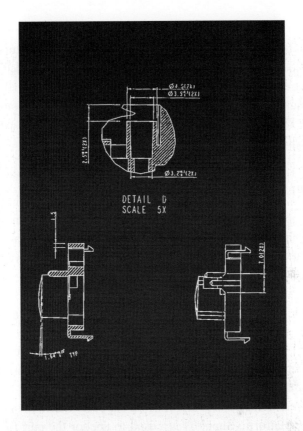

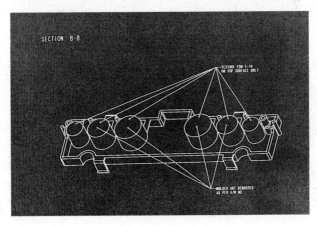

Figure 8.10: Working drawing showing the sectional and isometric views of the Button Tree Display

locating pins and LED holder. The second part is the dimensions taken from the button set, and the dimensions of the catch fall into the third part. Related to the different parts, the results are sub-divided into main measurements (> 10 mm) and detailed measurements (≤ 10 mm). For both readings, the deviations from the actual reading are computed and tabulated. The deviations for both measurements are plotted and compared.

8.3.3.2   *Results of the Measurements*

The graphs plotted facilitate a detailed comparison concerning the five different techniques namely the SGC, LOM, SOUP, SCS and FDM, in terms of the main and detailed measurements and their deviations from the designed dimensions.

From visual inspections, none of the thin ribs or walls is missing. This shows that all the five processes are capable of making wall thickness as thin as 0.5 mm.

However one of the locating pins of the LOM test pieces is tilted and the catches of the SGC and FDM test pieces are missing. This is due to the catches being too weak and they came off when the supports were removed. SLA objects are built on supports rather than directly on the elevator platform. Supports are used to anchor the part firmly to the platform and prevent distortion during the building process. The SOUP, FDM and SCS test pieces came with supports, thus they had to be removed before any measurements were taken. The button sets of the five test pieces are slanted due to its weight and the weak hinges. Tables 8.9 and 8.10 list the measurements taken and their deviations from the designed nominal values.

The plots illustrated in Figures 8.11 and 8.12 show that SGC achieved better measurements or fewer deviations as compared to other processes in detailed measurements. For the main measurements, FDM attained lesser deviations. On the other hand, SOUP produced the higher deviations for both the detailed and main measurements. From all the measurements taken, the greater deviations are observed for the radii of the button set and the angles of the LED holder. These

Table 8.9: Main measurements (>10 mm) of the five benchmark test pieces

Drawing Dimensions	(mm)	Main Measurements										
		Measured Dimensions (mm)					Deviations (mm)				
		SGC	SOUP	LOM	SCS	FDM	SGC	SOUP	LOM	SCS	FDM
Frame, location pins and LED hold	128.0	128.3	129.1	128.2	128.7	128.7	0.3	1.1	0.2	0.7	0.7
	27.0	27.1	27.3	27.1	27.1	26.8	0.1	0.3	0.1	0.1	-0.2
	65.0	65.7	65.5	64	64.4	65.6	0.7	0.5	-1.0	-0.6	0.6
	47.5	47.6	47.5	48.7	47.9	48.7	0.1	0.0	1.2	0.4	1.2
	31.25	31.3	31.5	32	31.7	31.2	0.05	0.25	0.75	0.4	-0.05
Button-tree	R31.5	30.1	14.5	33.5	28.0	29.5	-1.4	-17.0	2.0	-3.5	-2.0
	R21.0	14.2	7.0	32.0	20.5	20.5	-6.8	-14.0	11.0	-0.5	-0.5
	R12.0	15.0	8.0	14.0	13.0	11.0	3.0	-4.0	2.0	1.0	-1.0
	16.0	16.1	16.1	15.9	16.0	16.0	0.1	0.1	-0.1	0.0	0.0
	13.0	13.0	13.1	12.9	13.0	12.7	0.0	0.1	-0.1	0.0	-0.3
	22.75	23.0	23.8	24.4	22.45	22.3	0.25	1.05	1.65	-0.3	-0.45
Catch	33.0	32.3	33.1	32.3	32.5	32.9	-0.7	0.1	-0.7	-0.5	-0.1
	17.5	17.4	17.0	17.6	17.8	17.4	-0.1	-0.5	0.1	0.3	-0.1

Table 8.10: Detailed measurements (= 10 mm) of the five benchmark test pieces

Drawing Dimensions (mm)		Detailed Measurements										
		Measured Dimensions (mm)					Deviations (mm)					
		SGC	SOUP	LOM	SCS	FDM	SGC	SOUP	LOM	SCS	FDM
Location pins	8.0	7.2	8.2	8.0	8.1	8.8	-0.8	0.2	0.0	0.1	0.8
	5.7	4.7	6.8	6.5	7.6	7.4	-1.0	-1.1	-0.8	-1.9	-1.7
	45	39.4	31.8	37	56	54	-5.6	-13.2	-8.0	11.0	9.0
Button-tree	8.25	8.8	8.0	8.4	8.0	7.9	0.55	-0.25	0.15	-0.25	-0.35
	9.75	10.1	10.2	10.7	10.1	10.0	0.35	0.45	0.95	0.35	0.25
	10.0	10.1	10.1	9.9	10.0	9.9	0.1	0.1	-0.1	0.0	-0.1
	5.2	4.9	4.8	4.5	3.6	3.7	-0.3	-0.4	-0.7	-1.6	-1.5
	0.5	0.5	1.0	0.8	0.5	0.4	0.0	0.5	0.3	0.0	-0.1
Catch	1.6	1.5	1.1	1.4	1.7	1.9	-0.1	-0.5	-0.2	0.1	0.3
	0.8	1.0	1.0	1.3	1.0	0.7	0.2	0.2	0.5	0.2	-0.1

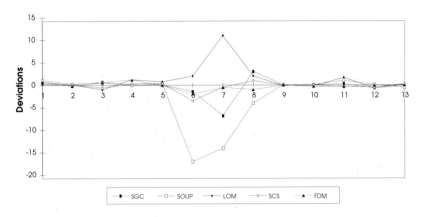

Figure 8.11: Deviation of main measurements (>10 mm) from the nominal values for the five benchmark test pieces

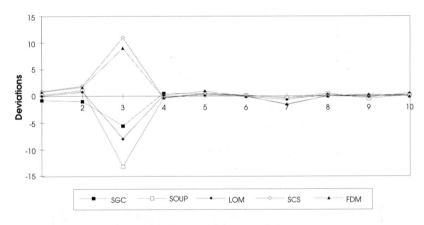

Figure 8.12: Deviation of detailed measurements (= 10 mm) from the nominal values for the five benchmark test pieces

indicated that RP systems have limitations in producing good curve surfaces.

The deviations could be due to the following errors in taking the measurements:

(1) The poor surface finishes of the test pieces resulted in not taking the actual dimensions.

(2) The test pieces are not strong as it deformed under pressure.
(3) The supports of the SOUP test piece were not properly removed.

8.3.4   Other Benchmarking Case Studies

Other than the benchmarking performed for Thompson Multimedia, there are also other benchmarking studies carried out by several researchers and Table 8.11(a)–(e) lists some of the significant ones from 1994 to 1996 [5–19].

Table 8.11(a): Benchmarking studies published in 1999

1.	Hardro, Peter J. & Stucker, Brent	Die Cast Tooling from Rapid Prototyping	Rapid Prototyping & Manufacturing '99, Rosemont, Illinois
	A "real-life" benchmark tool was chosen and has been built in RapidSteel 2.0 and compared to a conventionally made standard H13 steel tool of the same geometry.		
2.	Farentinos, S. & Khoshnevis, B.	A Study of Accuracy in Automated Fabrication	http://www.pmli.com/ accuray_study.html
	SLS, LOM, FDM and five different SLA systems and build styles were used. Dimensional accuracy was measured for 102 locations per sample. The test part was solely designed for the purpose of the study.		
3.	Shellabear, Mike	Benchmark Study of Accuracy and Surface Quality in RP Models	Task 4.2 Report 2 of Brite-EuRam project BE-2051: Process chains for Rapid Technical Prototypes (RAPTEC)
	In this study, 12 different systems were used including the most common ones and also KIRA PLT-A3, 3D systems Actua, Sanders MM6 and Z-Crop. Z402. Together with 22 different materials and in some cases varying build styles, this gave a total number of 44 combinations of build samples. Dimensional accuracy and surface finish were measured and compared. The test part has a special test geometry.		

Table 8.11(b): Benchmarking studies published in 1998

4.	Roberts, S.D. & Illston, T.J.	Direct Rapid Prototype Injection Molding Tools	Proceedings of 7th European Conference on Rapid Prototyping and Manufacturing, Aachen
	SLS, LOM, SGC and four different SLA systems were represented with one material each. A few main dimensions were measured on every male and female insert. Injection molding data and performance were recorded. The test part was a rather small and simple handwheel from Rover car company.		
5.	Loose, K. & Nakagawa, T.	Benchmarking Various Methods of Layer Manufacturing Systems in Rapid Prototyping	
	A benchmark part was designed for the test and built on 15 commercially available RP machines including some Japanese systems. Dimensions were measured and building time was noted. The processes were evaluated from a concept modeling point of view, i.e., with regard to the dimensional part accuracy, and the total manufacturing and systems list prices.		

Table 8.11(c): Benchmarking studies published in 1996

6.	Paul, B.K. & Barkaran, S.	Issues in fabricating manufacturing tooling using powder-based addictive freeform fabrication	Journal of Materials Processing Technology V 61 (1996) pp. 168–172
	A review of different approaches to direct fabrication of tooling using powder-based varieties of the materials on SLA, FDM, LOM, SLS, and FPM systems. Some accuracy data and material properties are given.		
7.	Reeves, P.E. & Cobb, R.C.	Surface Deviation Modeling of LMT Processes — a Comparative Analysis	Proceedings of 5th European Conference on Rapid Prototyping and Manufacturing, Helsinki
	This paper discusses how mathematical analysis of the surface roughness measurements of RP component could be used to derive the optimum build direction or to plan post process finishing of the parts. Six RP systems and some variation in build styles lead to nine different combinations of samples. Some data about surface roughness, surface flatness and key dimensions for the test parts are given. A specially developed benchmark part is used.		

Table 8.11(c): (*Continued*)

8.	Carosi, A., Pocci, D., Iuliano, L., & Settiner, L.	Investigation on Stereolithography Accuracy on both Solid & Quickcast Parts	Proceedings of 5th European Conference on Rapid Prototyping and Manufacturing, Helsinki
	A SLA machine was used to build the 3D System Benchmark Part in both solid and Quickcast build styles, and with varying process parameters. 89 measurements of dimensions and surface roughness were taken and the results were analyzed mathematically.		
9.	Almond, H.	Investment Casting Tooling Project	EARP (European Action on Rapid Prototyping) newsletter no 9
	Models were produced from all the major RP systems and in some cases from more than one manufacturer in the materials most suited to investment casting. A large number of dimensional measurements were made. The test parts were tool inserts for injection molding of a small plastics part.		

Table 8.11(d): Benchmarking studies published in 1995

10.	Ippolito, R., Iiliano, L. & Gatto, A.	Benchmarking of Rapid Prototyping Techniques in Terms of Dimensional Accuracy and Surface Finish	Annals of the CIRP, vol 44/1/1995
	An extended version of the project described in the report above. The benchmark component from 3D Machine Crop. was built once on a SLA, a SGC, and a LOM system, respectively, and twice on a FDM and a DTM system in two different materials. Accuracy and surface roughness measurements were reported and micrographs included.		
11.	Dickens, P.M. *et al.*	Conversion of RP models to investment castings	Rapid Prototyping Journals Vol. 1, number 4, 1995
	The six RP systems participated, each with the materials and build style most suitable for investment casting. Accuracy was measured for almost 20 dimensions and the surface finish was assessed. The test part was a component of a window-wiper mechanism from a German carmaker.		

Table 8.11(e): Benchmarking studies published in 1994

12.	Aubin, R.F.	A World Wide Assessment of Rapid Prototyping Technologies	UTRC Report No. 94-13 initiated by Intelligent Manufacturing Systems IMS Test Case on Rapid Product Development
	Most available systems and materials were included. A very complete array of handling, technological and practical aspects was covered. The test parts were specially developed combinations of different form features.		
13.	Schmidt, L.	A Benchmarking Comparison of Commercial Techniques in Rapid Prototyping Systems	Proceedings of Rapid Prototyping and Manufacturing Conference '94, Dearborn
	Most available systems were included in this benchmark. The main focus is on build times including pre- and post-processing and on costs. The test part was a small speedometer adapter — "Chrysler part".		
14.	Juster, N.P. & Childs, T.H.C.	A Comparison of Rapid Prototyping Processes	Proceedings of 3th European Conference on Rapid Prototyping and Manufacturing, Nottingham
	A specially designed benchmark component was built on four SLA systems and SL, FDM, and LOM systems. 112 measurements were taken. Only accuracy results are reported.		
15.	Ippolito, R., Iiliano, L. & de Fhilippi, A.	A New User Part for Performances Evaluation of Rapid Prototyping Systems	Proceedings of 3th European Conference on Rapid Prototyping and Manufacturing, Nottingham
	The specially designed benchmark component originally introduced by 3D Machine Corp. was built on SLA, SGC, FDM, and LOM systems. 210 measurements were taken. Only accuracy results are reported. A new benchmark part was proposed and built on a SLA and a SGC system. Accuracy and surface roughness values are reported.		

8.4   INDUSTRIAL GROWTH

The RP industry has enjoyed tremendous growth since the first system was introduced in 1988. The rate of growth has also been significant. Right up to 1999 the industry was enjoying a two-digit growth rate annually. In 2001, the RP industry continues to expand, though not at anywhere near the same rate as before. More systems are installed, more materials for these systems are used, and more applications for the technology are uncovered [20]. However, the rate of growth has tapered off significantly since 2000. The events and economic conditions of 2001 had not helped to improve the situation.

The U.S. still dominates the production and sale of RP systems [20]. Nearly 80% of the systems sold in 2001 come from U.S. systems manufacturers, just like the previous three years. Japan comes in a distant second contributing about 11% of sales with Europe making up only 5% [20].

8.5   FURTHER DEVELOPMENT TRENDS

As the whole rapid prototyping industry moves forward, rapid prototyping-driven activities will continue to grow. The two likely areas that will drive this growth will be that of concept modeling and rapid manufacturing and tooling (RM & T) applications [21].

Concept modeling, rapid prototyping and rapid manufacturing and tooling represents the three application areas of all RP-driven activities. As RP-driven activities grow, they are likely to be led by concept modeling applications as well as rapid manufacturing and tooling applications. These and other developing trends observed in the RP industry will be discussed in the following sections.

8.5.1   Concept Modeling

Concept modeling can be seen as a special aspect of rapid prototyping in that concept modeling is used more for design reviews rather than for physically testing. The two most important considerations in choosing a concept modeler over a higher-end rapid prototyping systems are its speed and comparatively low cost. Although models built using concept modelers often lack accuracy and have poor resolution, many users still gladly accept them, as the models built are used mainly for visual assessment.

Concept modelers are significantly cheaper when compared to the high-end systems, approximately $50 000 for concept modelers and $300 000 for high-end rapid prototyping systems. In addition, users of concept modelers pay less for a part that fits within a 50 mm cube (this includes materials, machine depreciation, system maintenance and labor) versus a more expensive rate for a part built on, say a SLA 5000. Thus concept modelers could pose a notable challenge to companies marketing high-end rapid prototyping systems, especially when a tremendous amount of effort and resources are put in to perfect the RP technology.

Another reason for the increase in the use of concept modelers is their relatively shorter fabrication time. Parts built by high-end rapid prototyping machine can take up to a few days or more to deliver a part, depending on factors like part size, workload and efficiency of the operation. This is mainly due to the fact that many of these systems are often located at a central site in a company and away from the design stations. Time is wasted in sending the STL file from the design station to the site, scheduling the work and sending the parts back. Concept modelers, however, do not have such a problem, as they are normally small in size, clean, safe and have a quiet operation, which allows them to be installed in an office environment, near to end users; running next to a CAD workstation, fax machines or copier. Hence, distance and time no longer separate users and the machines and it is possible for designers to start and complete the concept prototype in a single day or two.

As the concept modeler has just started to take a foothold in the industry, there are still many negative aspects, like the lack in reliability or flaws in the systems, as customers who have used such systems had complained. Even though manufacturers have put in great efforts to work on these problems and many have been remedied, many customers are still skeptical about their effectiveness and performance.

Nevertheless, with improved technology and design, concept modelers still have the potential for a bright future.

8.5.2   Rapid Manufacturing and Tooling

Rapid manufacturing and tooling (RM & T) has evolved from the application, rapid prototyping in manufacturing. It refers to a process that uses Rapid Prototyping technology to produce finished manufactured parts or templates, which are then used in manufacturing processes like molding or casting, directly building a limited volume of small prototypes. However, it is highly unlikely that this RP-driven process will ever reach the kind of production capacity of processes such as injection molding [21].

Typically, the costs incurred using the RM & T application is much lower than the conventional tooling cost. The tooling time is also notably lesser than the conventional tooling time. This is mainly due to the fact that rapid prototyping processes are able to fabricate models directly and quickly without the need of going through various stages of the conventional manufacturing process (i.e., post-processing). Moreover, it is more economical to build small intricate parts in low quantities using RM & T as there is compelling evidence that rapid manufacturing may be less expensive than traditional manufacturing approaches [22].

In addition, RM & T has the ability to make changes to design at low cost even after the product has reached the production stage. Such a feature is highly desirable as it is usually very expensive to make modifications to the molds (if the molding process is the fabrication for the finished product) and the time needed to modify the mold will also likely be lengthy. Thus RM & T is much preferred over the conventional tooling method in the early stages of the manufacturing cycle as modifications can be done almost instantaneously by simply editing the STL files and the cost incurred is relatively low.

However, the downside to rapid manufacturing processes is its lack of an appropriate range of materials for this application. Not all materials available are suitable for all products but nevertheless, current materials such as sintered nylon, epoxy resin and composite materials are often good enough for most low volume applications. Moreover, since RM & T has been successfully implemented in the aerospace industry, which imposes stringent quality demands, it is entirely possible that such applications can also be effectively implemented in other industries.

8.5.3 Others

Beside the two prominent development trends discussed, there are also other trends observed in the development of RP.

8.5.3.1 *Ease of use*

One such development trends is to make the rapid prototyping systems easier to use. Many RP developers are focusing their efforts in creating system software interfaces that are more familiar to users and also have the capability to ease the users' burden in operating the system. Much software developed is designed, based on the familiar and friendlier Microsoft's Windows interface rather than the Disk Operating System (DOS) interface. These software also readily accept CAD files with commonly used extensions like STL, DXF, SLC, etc. Moreover, some software, like Stratasys' Insight allows users to drag and drop CAD files to start modeling and provides a build log to track the status of individual jobs and the time to completion. In addition, developers also try to incorporate automatic features so as to make it easier to operate the rapid prototyping systems.

8.5.3.2 *Standardization of Systems*

Another development trend is the development towards standardization in the rapid prototyping industry. When the first rapid prototyping system was commercially launched in the late 80's by 3D Systems, many companies in the industry started to implement their own standards, like different CAD file inputs. However, as the industry advances, many companies have settled for several more commonly used CAD file inputs like STL and SLC formats. A current attempt for an international RP CAD file input standard is the evolving ISO 10303 under the auspices of the International Standards of Organization.

8.5.3.3   *Improvement of Building Speed and Prototype Quality*

One other development trend is the improvement of the building speed and prototype quality of rapid prototyping systems. Newer systems introduced are rapidly replacing older systems that are slow in building time and weak in accuracy. When Stratasys introduced FDM Maxum, an improved version of its older system, FDM Quantum, it was able to operate 50% [23] faster and was able to build layers from as thin as 0.178 mm [24]. Similarly, Solidscape introduced PatternMaster™, which is capable of operating two times as fast as its other model, ModelMaker II.

8.5.3.4   *Ease of Post-Processing*

Making post-processing easier to carry out or attempts to eliminate it altogether is another notable development trend. As systems are able to produce parts with high dimensional accuracy and better surface finish, the need for elaborate post-processing to be carried out has also diminished. In addition, some companies like Stratasys introduced new technology like water-soluble supports, WaterWorks™, which requires users to simply immerse the prototype into a water-based solution and wash away the supports. Hence, it eliminates the tediousness of spending several hours to remove supports from newly created prototypes and cleaning them afterwards.

8.5.3.5   *Improvement of Material Properties*

Another development trend is the effort towards improvement of material properties. In the earlier days of RP development, due mainly to the limitations of the technology, models built were often poor in quality and lacked desired physical and mechanical properties. Thus they were used mainly for visualization purposes in the early design stages. Even so, they do not work well as concept modelers. Today, many researchers are experimenting with a wider range of materials that are suitable for the intended applications, like rapid tooling, and to improve on the physical and mechanical properties of the fabricated models. SLS, for example, offers nylon and glass-filled nylon powders to produced toughen parts, two different sand-based materials for making sand casting molds and a steel-based powder called LaserForm ST-100 for making sturdier metal injection molds.

REFERENCES

[1] Wohlers, T., *Wohlers Report 2001: Rapid Prototyping & Tooling State of the Industry*, Wohlers Association Inc., 2001.

[2] Economic Development Board, Singapore, "Report on census of industrial production," Research and Statistics Unit, Economic Development Board, Singapore, 1993.

[3] Wohlers, T., "Rapid prototyping growth continues," *Rapid Prototyping Report* **4**(6) (1994): 6–8.

[4] Wall, M.B., "Making sense of prototyping technologies for product design," MSc. Thesis, MIT, USA, 1991.

[5] Hardro, P.J. and Stucker, B., *Die Cast Tooling from Rapid Prototyping. Rapid Prototyping & Manufacturing*, Rosemont, Illinois, 1999.

[6] Farentinos, S. and Khoshnevis, B., "A study of accuracy in automated fabrication," URL: http://www.pmli.com/accuracy_study.html/, August 1999.

[7] Shellabear, M., Benchmark Study of Accuracy and Surface Quality in RP Models, Task 4.2 Report 2 of Brite-EuRam project BE-2051: Process Chains for Rapid Technical Prototypes (RAPTEC), 1999.

[8] Roberts, S.D. and Illston, T.J., "Direct rapid prototype injection molding tools," *Proceedings of 7th European Conference on Rapid Prototyping and Manufacturing*, Aachen, 1998.

[9] Loose, K. and Nakagawa, T., *Benchmarking Various Methods of Layer Manufacturing Systems in Rapid Prototyping*, Institute of Physical and Chemical Research (RIKEN) and Institute of Industrial Science, University of Tokyo, 1998.

[10] Paul, B.K. and Barkaran, S., "Issues in fabricating manufacturing tooling using powder-based addictive freeform fabrication," *Journal of Materials Processing Technology* **61** (1996): 168–172.

[11] Reeves, P.E. and Cobb, R.C., "Surface deviation modeling of LMT processes — A comparative analysis," *Proceedings of 5th European Conference on Rapid Prototyping and Manufacturing*, Helsinki, 1996.

[12] Carosi, A., Pocci, D., Iuliano, L., and Settiner, L., "Investigation on stereolithography accuracy on both solid and Quickcast parts," *Proceedings of 5th European Conference on Rapid Prototyping and Manufacturing*, Helsinki, 1996.

[13] Almond, H., Investment Casting Tooling Project. European Action on Rapid Prototyping, 9, 1996.

[14] Ippolito, R., Iiliano, L., and Gatto, A., "Benchmarking of rapid prototyping techniques in terms of dimensional accuracy and surface finish," *Annals of the CIRP* **44**(1) (1995).

[15] Dickens, P.M., "Conversion of RP models to investment castings," *Rapid Prototyping Journals* **1**(4) (1995).

[16] Aubin, R.F., A World Wide Assessment of Rapid Prototyping Technologies. UTRC Report No. 94-13 initiated by Intelligent Manufacturing Systems IMS Test Case on Rapid Product Development, 1994.

[17] Schmidt, L., "A benchmarking comparison of commercial techniques in rapid prototyping systems," *Proceedings of Rapid Prototyping and Manufacturing Conference '94*, Dearborn, 1994.

[18] Juster, N.P. and Childs, T.H.C., "A comparison of rapid prototyping processes," *Proceedings of 3th European Conference on Rapid Prototyping and Manufacturing*, Nottingham, 1994.

[19] Ippolito, R., Iiliano, L., and de Fhilippi, A., "A new user part for performances evaluation of rapid prototyping systems," *Proceedings of 3th European Conference on Rapid Prototyping and Manufacturing*, Nottingham, 1994.

[20] Wohlers, T., *Wohlers Report 2002: Rapid Prototyping & Tooling State of the Industry Executive Summary*, Wohlers Association Inc., 2002.

[21] Wohlers, T., *Wohlers Report 2000: Rapid Prototyping & Tooling State of the Industry*, Wohlers Association Inc., 2000.

[22] Wuensche, R., Using Direct Croning to Decrease the Cost of Small Batch Production. Engine Technology International 2000 Annual Showcase Review for ACTech GmbH, 2000.

[23] Rapid Prototyping Report, *Stratasys Prodigy and Maxum* **11**(4), CAD/CAM Publishing Inc., April 2001: 5.

[24] Stratasys Inc., FDM Maxum System Specification. URL: http://www.stratasys.com/, July 2001.

PROBLEMS

1. What are the considerations when choosing a service bureau?

2. What are the components that make up the total cost of a part built by a service bureau?

3. What are the major considerations in assessing the economic feasibility of a proposed RP service bureau?

4. Is there a correlation between the type of RP systems and the industry in which the prototypes are used? Why?

5. Describe the marketing strategies a service bureau can employ.

6. Given that the total revenue is US$100 000, total expenses is US$70 000, corporate tax is 30% and total capital investment is US$200 000, calculate the rate of return. Assuming a 15% minimum acceptable rate of return, would you proceed with the venture?

7. Name the reasons for benchmarking.

8. Describe in detail the RP benchmarking methodology.

9. In what form of material can Rapid Prototyping Systems be classified as solid-based? Name three such systems.

10. How have the primary and secondary markets for RP performed over the years?

11. What are the likely trends in RP systems growth?

Appendix
LIST OF RP COMPANIES

3D Systems Inc. (SLA, MJM and SLS)
26081 Avenue Hall
Valencia, CA 91355
USA
US toll free phone: 888-337-9786
Phone: 661-295-5600
Fax: 661-294-8406
Email: moreinfo@3dsystems.com
URL: www.3dsystems.com

Contact Person: Lee Dockstader
Email: DockstadaL@3DSystems.Com

Aeromet Corporation (LasForm)
7623 Anagram Drive
Eden Prairie, Minnesota 55344
USA
Phone: 952-974-1800
Fax: 952-974-1801
Email: info@aerometcorp.com
URL: www.aerometcorp.com

Arcam AB (EBM)
Krokslätts Fabriker 30
SE-431 37 Mölndal
Sweden
Phone: 46-31-710-32-00
Fax: 46-31-710-32-01
Email: info@arcam.com
URL: www.arcam.com

Autostrade Co. Ltd. (E-DARTS)
13-54 Ueno-machi
Oita-City
Oita 087-0832
Japan
Phone: 81-97-543-1491
Fax: 81-97-545-3910
Email: rps@autostrade.co.jp
URL: www.autostrade.co.jp

Contact Person: Toshiyuki Akamine (President of Autostrade Co. Ltd.)
Email: akamine@autostrade.co.jp

Beijing Yinhua Rapid Prototypes Making and
Mould Technology Co. Ltd. (SSM, MEM and M-RPM)
Department of Mechanical Engineering
Tsinghua University
Beijing 100084
China
Phone: 8610-6278-2988
Fax: 8610-6278-5718
URL: www.geocities.com/CollegePark/Lab/8600/clrfprod.htm

Contact Person: Professor Yan Xu Tao
Email: yanxt@mail.tsinghua.edu.cn

CAM-LEM Inc. (CL 100)
1768 E. 25th St.
Cleveland, OH 44114
USA
Phone: 216-391-7750
Fax: 216-579-9225
Email: info@camlem.com
URL: www.camlem.com/

Contact Person: Brian B. Mathewson (Engineer)
Email: bbm@camlem.com

CMET Inc. (SOUP)/Teijin Seiki (Soliform)
Kamata Tsukimura Bldg.
5-15-8, Kamata, Ohtaku
Tokyo 144-0052
Japan
Phone: 81-3-3739-6611
Fax: 81-3-3739-6680
URL: www.cmet.co.jp

Contact Person: HAGIWARA, Tsuneo Ph.D. (Executive Director)
Email: hagi@cmet.co.jp

Cubic Technologies Inc. (LOM)
1000E. Domiguez Street
Carsion, California 90746-3608
USA
Tel: 310-965-0006
Fax: 310-965-0141
Email: info@CubicTechnologies.com
URL: www.cubictechnologies.com

Cubital Ltd. (SGC)
12 Habonim St.
Ramat-Gan 52462
Israel
Phone: 972-3-751-7945
Fax: 972-3-575-0421
Email: cubital@cubital.com
URL: www.cubital.com

Contact Person: Mr Eitan

D-MEC Ltd. (SCS)
JSR Building
2-11-24 Tsukiji
Chuo-ku, Tokyo 104-0045
Japan
Phone: 81-3-5565-6661
Fax: 81-3-5565-6641
Email: tokyo@d-mec.co.jp
URL: www.d-mec.co.jp

Ennex Corporation (Offset Fabbers)
11465 Washington Place
Los Angeles, California 90066
USA
Phone: 310-397-1314
Fax: 310-397-8539
Email: fabbers@Ennex.com
URL: www.Ennex.com

EOS GmbH (EOSINT)
Robert-Striling-Ring 1
D-82152 Krailing
Germany
Phone : 49-0-89-893-36-0 (with effect from 16.09.2002)
Email: info@eos-gmbh.de
URL: www.eos-gmbh.de

Extrude Hone (ProMetal)
1 Industry Blvd.
P.O. Box 1000
Irwin, PA. 15642
USA
Toll Free: 1-800-367-1109
Phone: 724-863-5900
Fax: 724-863-8759
Email: exhone@extrudehone.com
URL: www.extrudehone.com/

Fraunhofer-Gessellschaft (MJS)
Leonrodstraße 54
D-80636 München
Germany
Phone: 49-0-89/12-05-01
Fax: 49-0-89/12-05-317
Email: info@fraunhofer.de
URL: www.fraunhofer.de

Fraunhofer Representative Office Singapore
Dr. Dagmar Martin-Vosshage
15 Beach Rd., #05-08
Beach Centre
Singapore 18 96 77
Phone: 65-6338-4355
Fax: 65-6338-9456
Email: fhgsin@singnet.com.sg
URL: www.fhg.de/fhgsin-e.html

Generis GmbH (GS)
Am Mittleren Moos 15
D-86167 Augsburg
Germany
Phone: 49-0-821/74-83-100
Fax: 49-0-821/74-83-111
Email: info@generis.de
URL: www.generis.de or www.generis-systems.com

Kira Corporation Ltd. (PLT)
Tomiyoshishinden, Kira-Cho
Hazu-gun, Aichi
Japan
Phone: 81-563-32-1161
Fax: 81-563-32-3241
Email: info@kiracorp.co.jp
URL: www.kiracorp.co.jp

Meiko Co. Ltd.
732 Shimoimai, Futaba-cho, Kitakoma-gun
Yamanashi-ken, 407-0105
Japan
Phone: 81-551-28-5111
Fax: 81-551-28-5121
Email: info@meiko-inc.co.jp
URL: www.meiko-inc.co.jp (only in Japanese)

Contact Person: Takuya Akiyama (System Department)
Email: t-akiyama@meiko-inc.co.jp

Optomec Inc. (LENS)
3911 Singer Blvd. NE
Albuquerque, NM 87109
USA
Phone: 505-761-8250
Fax: 505-761-6638
Email: customerservice@optomec.com
URL: www.optomec.com

Contact Person: Theresa Chavez-Romero (Customer Service Manager)
Email: tchavez-romero@optomec.com

POM Inc. (DMD)
2350 Pontiac Road
Auburn Hills, Michigan 48326
USA
Phone: 248-409-7900
Fax: 248-409-7901
Email: info@pom.net
URL: www.pom.net/

Solidscape Inc. (ModelMaker and PatternMaster)
316 Daniel Webster Highway
Merrimack, New Hampshire 03054-4115
USA
Phone: 603-429-9700
Fax: 603-424-1850
Email: precision@solid-scape.com
URL: www.solid-scape.com

Soligen Inc. (DSPC)
19408 Londelius St.
Northridge, CA 91324
USA
Phone: 818-718-1221
Fax: 818-718-0760
URL: www.soligen.com

Contact Person: Gary Kanegis (Vice President, Sales)
Email: gary@soligen.com

Stratasys Inc. (FDM)
14950 Martin Drive, Eden Prairie
MN 55344-2020
USA
Toll Free: 888-480-3548
Phone: 1-952-937-3000
Fax: 1-952-937-0070
Email: info@stratasys.com
URL: www.stratasys.com

Teijin Seiki (Soliform)
(See CMET)

Therics Inc. (TheriForm)
115 Campus Drive
Princeton, NJ 08540
USA
Phone: 609-514-7200
Fax: 609-514-7219
Email: info@therics.com
URL: www.therics.com/

Z Corporation (3DP)
20 North Avenue
Burlington, MA 01803
USA
Phone: 781-852-5005
Fax: 781-852-5100
Email: sales@zcorp.com
URL: www.zcorp.com

CD-ROM ATTACHMENT

Multimedia is a very effective tool in enhancing the learning experience. At the very least, it enables self-paced, self-controlled interactive learning using the media of visual graphics, sound, animation and text. To better introduce and illustrate the subject of *Rapid Prototyping*, an executable multimedia program has been encoded in a CD-ROM attachment that comes with the book. It serves as an important supplemental learning tool to better understand the principles and processes of RP.

More than twenty different commercial RP systems are described in Chapters 3–5. However, only the six most matured techniques and videos on their processes are demonstrated in the CD-ROM. These six techniques and the length of their videos are listed below in Table I:

Table I: The six RP techniques and their corresponding video lengths

Technique	Movie Length/min
Stereolithography Apparatus (SLA)	1:17
Solid Ground Curing (SGC)	2:44
Laminated Object Manufacturing (LOM)	4:30
Fused Deposition Modeling (FDM)	2:07
Selective Laser Sintering (SLS)	1:17
Three-Dimensional Printing (3DP)	3:40

The working mechanisms of these six methods are interestingly different from one another. In addition, the CD-ROM also includes a basic introduction, RP processes chain, RP data formats, applications and benchmarking. While the book describes the principles and illustrates with diagrams the working principle, the CD-ROM goes a step further and shows the working mechanism in motion. Animation techniques through the use of Macromedia Flash enhance understanding through graphical illustration. The integration of various media (e.g., graphics, sound, animation and text) is done using Macromedia Director. Additional information on each of the techniques such as product information, application areas and key advantages, makes the CD-ROM a complete computer-aided self-learning software about RP systems. Together with the book, the CD-ROM will also provide a directly useful aid to lecturers and trainers in the teaching of the subject of Rapid Prototyping.

CD-ROM USER GUIDE

This user guide will provide the information needed to run and use the program smoothly.

System Requirements

The system requirements define the minimum configurations that your system needs in order to run the multimedia package smoothly. In some cases, not meeting this minimum requirement may also allow the program to run. However, it is recommended that the minimum requirements be met in order to fully benefit from this multimedia course package.

The system requirements are:

- IBM Pentium PC or 100% compatible with 8 MB of RAM
- Graphics card with VGA resolution or higher
- Windows 95 or higher
- 600 MB of hard disk space (if program is installed into the hard disk)
- Sound card and speakers (Recommended)

For trouble-free running or installation of the program, do ensure that the computer system using this program meets the minimum system requirements as stated above.

Before Running or Installing RP Multimedia CD-ROM

The next section provides some simple steps to help prepare you to install or run the RP Multimedia CD-ROM. Before starting, please read the License Agreement provided at the end of this user guide. When you install or run the program on the CD, you agree to comply with the agreement.

Installing RP Multimedia CD-ROM

Although the CD-ROM can be run without installation into the hard disk, the performance of the multimedia package will be lower than when it is installed onto the hard disk. Therefore, it is recommended that you install this courseware onto the hard disk in order to maximize the benefits of the multimedia experience.

Windows 95 (or higher) Installation

To install the RP Multimedia package onto the hard disk, follow the steps below:

(1) Start Windows 95 or higher by turning on your computer.
(2) Insert the RP Multimedia CD into your CD-ROM drive.
(3) Click on the "Start" menu button and select "Run".
(4) In the "Run" dialog box, type "d:\setup" and click on the "Ok" button. If your CD-ROM drive is not drive "d", substitute its drive letter in place of "d".

Running the Program

The RP multimedia package can be run directly from the CD or from the hard disk. Either way, the multimedia package will remain the same. As the hard disk is generally a faster device than a CD-ROM drive, it will therefore provide better performance when running this program. Hence, it is recommended that the program be installed on the hard disk.

Running from the CD-ROM Drive

To run the program directly from the CD-ROM drive, follow the steps below:

(1) Start Windows 95 or higher by turning on your computer.
(2) Insert the RP Multimedia CD into your CD-ROM drive.
(3) Click on the "Start" menu button and select "Run".

(4) In the "Run" dialog box, type "d:\rpstart" and click on the "Ok" button. If your CD-ROM drive is not drive "d", substitute its drive letter in place of "d".

(5) The program will start.

Running from the Hard Disk

To run the program from the hard disk, ensure that the program is installed onto the hard disk. The RP multimedia CD is not required to run the program if it is already installed onto the hard disk. Then follow the steps below:

(1) Start Windows 95 or higher by turning on your computer.

(2) Click on the "Start" menu button and select "Run".

(3) In the "Run" dialog box, type "d:/projectorintro" and click on the "Ok" button. If your CD-ROM drive is not drive "d", substitute its drive letter in place of "d".

Getting Around the Program

The RP multimedia program is a courseware designed for students, lecturers and any person interested in learning more about RP. The user interface has been designed to be very simple and intuitive so that it will be easy even for first time users to understand how to move around the chapters and sub topics in the program.

Nevertheless, the next section will guide you through the various screens in the courseware and explain the functions of the icons on the screens in detail.

Main Introduction Screen

After the program has been started, a short introduction movie will be played. To skip this introduction movie, simply click on the "Skip" button. After the movie has ended, you will be presented with the Main Introduction screen or homepage.

Choosing a Chapter

To learn more about a particular chapter, move and point your mouse cursor to the word of that chapter in the homepage. The word will be lighted up to indicate the particular chapter is currently being selected. Clicking once on the left mouse key when a word is lighted up will bring you into that chapter.

Main Learning Screen

The main learning screen for every chapter will be basically the same and will look like Figure I. More advance interactive features of this program will be covered in a later section "Exploring the Interactive Features".

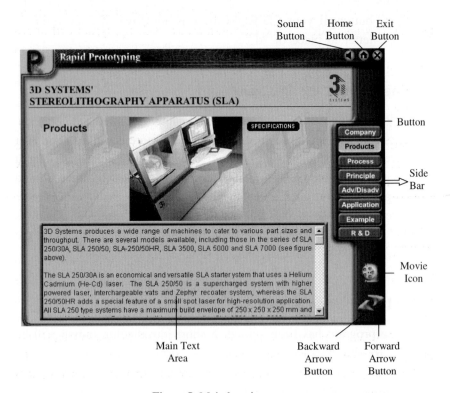

Figure I: Main learning screen

Components of the Main Learning Screen

The main learning screen comprises of the main text area and various icons as shown in the screen shot above. The main text area is where the text information of a particular chapter and sub topics will be displayed. It is also the area where the user will do most of the learning.

The icons are explained below:

(1) *Sound Button.* Click on the sound button icon to toggle the music on or off. If the music is playing, clicking on the icon will make the music stop playing. Clicking on the icon when the music is not playing, will start the music again.

(2) *Home Button.* Click on the home button to bring up the main introduction screen or homepage. In the homepage, choose a chapter to "jump" to and read more about that chapter's contents.

(3) *Exit Button.* Click on the exit button to end this program and return to Windows. When the exit button is clicked, a confirmation screen will appear. To confirm and exit the program, click on the "Yes" button. To cancel the exit program request, click on the "No" button.

(4) *Backward and Forward Arrow Buttons.* To advance a page forward, left mouse click on the forward arrow button once. Similarly, to go back to the previous page, left mouse click on the backward arrow button once.

Exploring the Interactive Features

The RP multimedia courseware is an interactive, fun and lively way to learn about RP and how they work. There are many interactive features that can be found in the program. Interactive features such as animations, graphics, tables and movies are "scattered" around the courseware to help enhance learning of those topics and concepts. Do enjoy and have fun exploring around!

The next section will highlight some of the interactive features that can be found within the courseware and how to activate them.

Mouse Over a Hot Spot

There are many "Hot spots" within the main text area in the learning screen. By moving the mouse cursor over such "hot spots", hidden text, graphics or animations will appear (or move) on the screen.

When the mouse cursor moves over the "Data Preparation" word, the text explaining data preparation appears to the right. The same goes for the "Mask Generation" and "Model Making" words.

This is one of the examples of the mouse over a "hot spot". There are many different "hot spots" in the entire courseware. So do explore and discover the hidden "hot spots" to help in understanding the concepts of RP better and faster.

Buttons

In each of the six RP techniques, there is a side bar for navigation within that section. Mouse over the side bar buttons and they turn red. Click on a side bar button and it will bring up the respective page. Figure I shows the side bar at the "Product" page of the SLA technique. Another type of button used in the RP multimedia is shown in Figure I. Click on the button to bring up the respective page.

Playing a Movie

There are six movies in this courseware to help explain the concepts of the RP technologies. Whenever a particular sub-topic or page contains a related movie, the movie icon will appear on the right of the main learning screen as seen in Figure I. Clicking on the movie icon will bring out the movie player and plays the movie.

Movie Player Controls:

(1) *Replay Button.* Restart the movie and playback from the beginning.
(2) *Pause Button.* Freeze the playing movie at a particular point. Pressing the "Pause" button again will resume the movie playback from that point.

(3) *Done Button*. End the movie and return to the previous page.

(4) *Sound Button*. The sound button works the same as that found in the main learning screen. Click on the sound button icon to toggle the music on or off.

INDEX

U.W.E.L. LEARNING RESOURCES

U.W.E.L. LEARNING RESOURCES